Plant Abiotic Stress Responses and Tolerance Mechanisms

Edited by Saddam Hussain, Tahir Hussain Awan, Ejaz Ahmad Waraich and Masood Iqbal Awan

Published in London, United Kingdom

Plant Abiotic Stress Responses and Tolerance Mechanisms
http://dx.doi.org/10.5772/intechopen.102138
Edited by Saddam Hussain, Tahir Hussain Awan, Ejaz Ahmad Waraich and Masood Iqbal Awan

Contributors
Parul Chowdhury, Diksha Vaishnav, Noreen Fatima, Faiza Khan, Asif Saeed, Tamer Gümüş, Sinan Meriç, Alp Ayan, Çimen Atak, Ahlem Zrig, Safia Belhadj, Taieb Tounekti, Habib Khemira, Salma Yousif Sidahmed Elsheikh, Megha Nisha Singh, Veena Devi, Amanpreet Kaur, Mehak Sethi, Gosangi Avinash, Saddam Hussain, Attiqa Rahman, Ghadeer M. Albadrani, Ejaz Ahmad Waraich, Tahir Hussain Awan, İlkay Yavaş

First published in London, United Kingdom, 2023 by IntechOpen
IntechOpen is the global imprint of INTECHOPEN LIMITED, registered in England and Wales, registration number: 11086078, 5 Princes Gate Court, London, SW7 2QJ, United Kingdom

British Library Cataloguing-in-Publication Data
A catalogue record for this book is available from the British Library

Additional hard and PDF copies can be obtained from orders@intechopen.com

Plant Abiotic Stress Responses and Tolerance Mechanisms
Edited by Saddam Hussain, Tahir Hussain Awan, Ejaz Ahmad Waraich and Masood Iqbal Awan
p. cm.
Print ISBN 978-1-80356-956-7
Online ISBN 978-1-80356-957-4
eBook (PDF) ISBN 978-1-80356-958-1

Meet the editors

Dr. Saddam Hussain is an Assistant Professor of Plant Stress Physiology at the Department of Agronomy, University of Agriculture Faisalabad (UAF), Pakistan. He has been recognized with several international distinctions and awards in recent years. He has published more than 180 refereed journal papers, many of which have sought to understand the morphological, physiological, biochemical, and molecular basis of crop responses to individual and concurrent abiotic stresses and mechanisms imparting tolerance for such stresses. He devised promising strategies for improving crop performance under sub-optimum conditions and suggested various novel indicators for augmenting stress tolerance in plants. Dr. Hussain was featured in the "Highly Cited Researchers" list by Clarivate/Web of Science in 2021 and 2022. During his professional career, he has availed several national/international grants as a principal investigator. He established strong research collaboration in several countries including China, the United States, Australia, Germany, Canada, and Kingdom of Saudia Arabia. To date, Dr. Hussain has supervised more than thirty postgraduate students as a major supervisor/ supervisory committee member. He serves as an editor/editorial board member of various international journals including the *Journal of Advanced Research*, *Ecotoxicology and Environmental Safety*, *Journal of Plant Growth Regulation*, *BMC Plant Biology*, *Food and Energy Security*, and *Crop Science-CSSA*, among others.

Dr. Ejaz Ahmad Waraich is an associate professor at the Department of Agronomy, University of Agriculture Faisalabad (UAF), Pakistan. He is the recipient of the 2011, 2012, and 2013 Research Productivity Award from the Pakistan Council for Science and Technology, the 2011 Endeavour Research Fellowship Award and 2015 Endeavour Executive Award from the Government of Australia, and a 2008–2009 post-doc fellowship award from HEC, Pakistan. He has published more than 140 refereed journal papers and has presented at several national and international conferences/symposia. He has supervised more than fifty postgraduate students as a major supervisor.

Dr. Tahir Hussain Awan is a senior scientist at Rice Research Institute, Kala Shah Kaku, Punjab, Pakistan. He obtained a BSc and MSc from the University of Agriculture, Faisalabad (UAF) in 1997 and 1999, respectively. He obtained a Ph.D. from the University of the Philippines Los Baños in 2015. He has more than 25 years of research experience in crop physiology and agronomy. He has published around 100 articles in various reputed journals. Dr. Awan's key research interests include stress physiology, weed biology and management, crop modeling, and eco-efficient cropping systems.

Dr. Masood Iqbal Awan has a background in crop physiology and agronomy. He is currently an assistant professor at the University of Agriculture Faisalabad (UAF), Pakistan. He obtained a BSc (2005) and MSc (2007) from UAF in 2005 and 2007, respectively. He obtained a Ph.D. from Wageningen University, the Netherlands in 2013. During 2015–2017, he was appointed as a university associate in the School of Land and Food, Tasmanian Institute of Agriculture (TIA), University of Tasmania (UTAS), Australia. His research focuses on cropping systems, specifically understanding how different elements (plants, soils, and people) interact and how to improve robustness, productivity, and sustainability for desirable outcomes. He has published papers in disciplinary and transdisciplinary journals. His key research interests include stress physiology, resource use efficiency, alternate production systems/crops, and eco-efficient cropping systems.

Contents

Preface

Climate changes have multifaceted effects on the consequences of abiotic stress, threatening the productivity and sustainability of agricultural systems. As sessile organisms, plants generally encounter a wide range of abiotic stresses such as salinity, drought, and temperature extremes during their life span. Under stress conditions, the survival of plants relies on the coordination of stress-adaptive physiological, metabolic, and structural alterations into endogenous developmental programs. Considering the current climate change, it is of great importance to understand the processes that underpin plant growth under stressful environments and design possible strategies to overcome the expected challenges. Moreover, understanding the plant tolerance mechanisms to different abiotic stresses is inevitable for developing stress-tolerant cultivars. *Plant Abiotic Stress Responses and Tolerance Mechanisms* provides the most recent understanding of plant morpho-physiological, biochemical, and molecular responses to different abiotic stresses and highlights the underlying tolerance mechanisms against these stressors. With contributions from internationally recognized scientists, this volume contains seven chapters covering topics such as current challenges and future threats of plant abiotic stresses, regulatory networks in plants under abiotic stresses, plant adaptation to temperature extremes and salinity, plant secondary metabolites and stress tolerance, and the role of phytohormones in plant stress tolerance.

I wish to express my gratitude to the contributing scientists for their overwhelming response and for readily accepting my invitation to contribute to this book. They not only shared their knowledge but also admirably integrated scattered information from diverse fields in composing the chapters and efficiently incorporated editorial suggestions. I greatly appreciate their commitment. I am also thankful to the IntechOpen team for their generous cooperation at every stage of the book production. I hope this volume will be useful to agriculturists, environmentalists, research scholars, students, and others working on plant abiotic stresses.

Dr. Saddam Hussain
Assistant Professor,
Department of Agronomy,
University of Agriculture,
Faisalabad, Pakistan
(Academic Editor)

Ejaz Ahmad Waraich
Associate Professor,
Department of Agronomy,
University of Agriculture,
Faisalabad, Pakistan

Tahir Hussain Awan
Senior Scientist,
Rice Research Institute,
Kala Shah Kaku, Punjab, Pakistan

Masood Iqbal Awan
Assistant Professor,
University of Agriculture,
Faisalabad, Sub-Campus Depalpur, Okara, Pakistan

Chapter 1

Perspective Chapter: Plant Abiotic Stress Factors – Current Challenges of Last Decades and Future Threats

Tamer Gümüş, Sinan Meriç, Alp Ayan and Çimen Atak

Abstract

All life forms, from the simplest to the most complicated, are inevitably exposed to altering environmental conditions in their natural habitats, gradually depending on their lifestyle. Unfavorable alterations drive these life forms either to avoidance or defense as a response. Most of the essential plant growth-promoting environmental factors can also turn out to be stress factors. Water as the most abundant molecule of all living cells can cause stress either in deficit as drought or in excess as waterlogging. Temperature is important for the maintenance of all biomolecules and metabolic reactions; hence, both low and high temperatures are deleterious stress factors. Even though the plants were exposed to various volcanic origin, heavy metals and pollutants and evolved molecular mechanisms during millions year of evolution, rapid urbanization, and industrial progress introduce brand new pollutants as micro- and nanoplastics as well as nanoparticles to plants like never before. This chapter defines and evaluates major environmental abiotic stress factors with an emphasis on the latest knowledge of molecular effects on plants. In addition, novel stress factors, such as nanoparticles and microplastics, are looked over as hot prospects for the future of plant abiotic stress areas.

Keywords: nanoparticles, microplastics, nanoplastics, abiotic stress tolerance

1. Introduction

Environment as a term originated from the French word “environ” by the meaning of encompass or surround. In the early twentieth century, a biologist, Jacob van Erkul, pioneered the subject to describe all the physical, chemical, and biological factors, which comprise and interdependence with living organisms. In a healthy environment, all living organisms borrow basic necessities of life from nature and deposit waste and pollutants as by-products. However, this deposit is in a rate of recycling capacities of nature itself. Following the industrial revolution in the late eighteenth to the middle of the nineteenth century, the transition from manufacturing methods based on hand production and manpower to fossil fuel-based steam power machine tools and mechanized factory production led to unprecedented use of natural sources and deposition of wastes and pollutants. Technological developments in textile manufacturing, iron industry, power production, chemicals, infrastructure, lighting,

IntechOpen

paper production, mining, transportation, and agriculture altered the social aspects of living standards, industrialization, consumer goods, urbanization, lifestyle, and labor conditions, which all led to rise of the population as never before [1].

Dry land environment referred to as "*terra firma*" was never hospitable to plants through their terrestrialization during over the billion years of evolution process from plastid-bearing algae lineages to the formation of desiccation-tolerant photosynthetic eukaryotes. Various algae lineages evolved to survive in terrestrial lifestyle after numerous biochemical and physiological photosynthetic adaptations in multiple independent times long before plants. However, land plants evolved from terrestrial life adapted unicellular filamentous freshwater algae, *Zygnematophycae*, in *Streptophyta* clade, once in 450 million years ago during the end of the Cambrian Era [2]. The first plants were seedless nonvascular plants with a basic low body plan called bryophyte. Evolution is a dynamic process and the first phase required adaptive traits, such as complex plant body plan, vascular tissues to conduct water and minerals to upper part and to provide stability, desiccation-tolerant seeds instead of spores, and stomata for gas exchange.

Drought is a major threat to agricultural food security today as it was to the first land plants during this evolution process. Water deficiency can be physically caused by insufficient rainfall, and poor storage capacity of soil based on its texture in arid and semiarid regions. It can also be caused physiologically when the plant transpiration rate surpasses the rate of water uptake. Water deficiency also has cellular effects, such as alterations in cytoplasmic solutes and water potential gradient, reduction in turgor pressure and wilting, shrinking of cell volume and membrane integrity, denaturation, and degradation of proteins and more. The intensity and the duration of the drought determine the effects on plants along with the genotype and the developmental stage of the plant as well as other environmental interactions [3]. In the present day, drought as the direct consequence of global warming brought secondary problems to the field such as salinity. Saline components naturally exist in rocky layers. Insufficient drainage of the surface salts from the plant root zone leads to accumulation. Poor irrigation strategies also add up to the concentration as water molecules evaporate and leave substances, such as chloride, sulfate, carbonate compounds of sodium, potassium, and magnesium. As drought, the complex effects of salinity on plants are mainly through osmotic properties. Imbalance in ion homeostasis, physiological water deficiency, and insufficient nutrient uptake is the most severe results. Excluding sodium and other saline components from roots prevents plants from the intake of 17 elements, which are essential to plant growth and development. Today, a phenomenon called nutrient stress is mostly associated with salinity and heavy metal stresses and leads to metabolic plant diseases, which reduces growth and yield significantly. Heavy metals, such as Cd (Cadmium), Hg (Mercury), and Pb (Lead), are toxic elements with no biological use for plants and other organisms, while Ni (Nickel), Cu (Copper), and Zn (Zinc) are double-edged blades that are micronutrients for plants in low concentrations and toxic in high. Plants developed strategies, such as cell wall binding, reduction of mobility through cell membrane, active efflux, compartmentalization, chelation, and sequestration with phytochelatins and metallothioneins. As in heavy metals, 17 essential nutrition elements are also nutritional stress factors in excess levels for plants.

In the second phase, former photosynthetic aquatic habitat organisms had to adapt to terrestrial dry land stressors, such as high irradiance of unfiltered sunlight, which leads to photooxidative damage through UV rays, severe heat, and cold as their plastids were not hard-wired to the cellular stress response pathways and hence, susceptible to stress-related photosynthetic gene expression inhibition. Plants are sessile organisms and are constantly exposed to high ambient temperatures through sun rays in various wavelengths. Combined with drought effects, internal heat may rise even to a higher degree as transpiration through stomata is reduced. Today, heat shock proteins (HSPs) are an important topic of interest as their expression significantly elevates when external body temperature increases from 5 to 10°C from the organism's regular growth temperature. They are also one of the most conserved protein families in all living cells. Another stress factor based on direct exposure to the sun is called light stress or photooxidative stress. As the absorbed photon energy surpasses the photosynthetic capacity of the photosystems in chloroplasts, excess energy forms reactive oxygen species (ROS), which may damage the photosystem and chloroplast, if not scavenged. Low (freezing) temperatures are also major threats to plants as seasonal temperature fluctuations are more frequent with global climate effects. Freezing mainly affects the membrane systems of plant cells and secondarily protein denaturation. Freezing tolerance is mainly associated with levels of metabolites, such as sugars, their respective enzyme activities, HSPs, lipids of cell membrane, abscisic acid, and other altered metabolism products, which depress water's freezing point in cells. In the evolution of plants, other adaptations as cell walls containing lignin-like monomers eased the dramatic effects of dehydration as well as provided stability and sturdiness to plants. The phenylpropanoid pathway contributed as a UV shield through phenolic compound production [4]. Modern-day plants facilitate phytohormone-mediated stress signaling during turbulent times. Abscisic acid, as one of the key phytohormones, increases under all osmotic stresses, such as drought, salinity, and freezing as well as under some other abiotic stresses as heavy metals. Even though it is called a stress hormone in plants, it also has important physiological duties on unstressed plants. Comparative genomic studies on ABA discovered 47 of 48 transcription factor (TF) families of modern land plants, which utilize complex gene expression regulatory networks, were also present in algal common ancestors, resulting in similar phytohormone-mediated signaling [5].

Undoubtfully, first land plants were forced to cope with various environmental abiotic stress factors in unfavorable conditions. Either adaptations or exaptations equipped plants during terrestrialization. Despite million years of evolutionary adaptations drought, temperature (heat and cold), nutrients, heavy metals, salinity, and more are still limiting abiotic stress factors in varying intensities and combinations for approximately 380.000 plant species, including a handful of agricultural crops among them [6]. Except for these common limiting stress factors, the persistence and steady increase of plastic pollution is an emerging global issue of the present day. Once plastics seep into the environment, they can take up to several hundred years to get degraded, persistently. The production rate of plastics has exceeded 350 million tons per year since the initiation of mass production in the 1950s [7]. Plastics escape into many different environments, from soil to the deep water of the oceans. It is estimated that 8.300 million metric tons (Mt) of raw plastic have been produced to date. As of 2015, approximately 6.300 Mt. of plastic waste was produced of which 9% was recycled, 12% incinerated, and 79% accumulated in landfills or in the natural environments. If current production and waste management trends continue, 12.000 Mt. of plastic waste will accumulate in landfills or in the natural environment by 2050 [8]. The dangerous effects of microplastics and nano plastics, also known as "next generation" organic pollutants, on the environment cause intense concerns. Previous studies have reported that plastic pollution affects the flora in terrestrial and aquatic ecosystems [9]. Plastics are new stressors that can be found around and in soil-dwelling organisms, especially plants. Plastics can degrade in different ways depending on the environmental conditions and the chemical composition of the plastic. These ways include mechanical, thermal, chemical, biological degradation, and light-induced photodegradation [10, 11]. In general, plastic polymer particles are divided into three main classes: macro plastics, which are particles larger than 5 mm, microplastics, which are particles between 5 mm and 100 nm, and nanoplastics, which are particles less than 100 nm [12]. Potential sources of microplastics include household and industrial waste, personal care products, city dust, road marking, marine coating, tire wear, and residues from the synthetic textile industry [13]. Microplastics are divided into two subclasses, such as primary and secondary microplastics, in terms of their origin. Primary microplastics have been created for direct use or as precursors to a variety of products in the preproduction of plastic pellets/nurdles, microbeads used in personal care products, industrial and cosmetic abrasives, exfoliants, and various consumer-use products. Secondary microplastics occur in the environment as a result of the breakdown of larger plastic debris. Secondary microplastics can leach into water bodies and are significantly involved in sewers and surface water [14]. Microplastics commonly detected in the environment are polyethylene (PE), polypropylene (PP), polystyrene (PS), polyvinyl chloride (PVC), polyamide (PA), and polyethylene terephthalate (PET). On the other hand, the term nanotechnology was first introduced to the literature by Taniguchi in 1974, as it was used for particles between 1 and 100 nm. When atoms are reduced to nano-size, they have a high surface-to-volume ratio that allows them to acquire new properties completely distinct from their macro-scale properties. For nanomaterials to have high reactivity, the surface area must be small, and the zeta potential must be as high [15]. Nanostructured materials obtained by nanoparticles are called nanomaterials. The change in the surface/volume ratio is obtained as a result of the change in the number of atoms on the material surface. Thanks to the change in atomic quantities, the surface/volume ratio changes. Changing the number of atoms on the surface of a material also affects surface energy. It has been stated in many studies that the effects of nanoparticles vary depending on the applied doses. As a result of exposure at high doses, it causes oxidative damage to biomolecules in plants and causes the formation of reactive oxygen derivatives. As a result of the damage caused by the formed ROS, damage that will result in cell death may occur. The root growth inhibition initiated by nanoparticle treatment in plants was associated with particular cell division errors and chromosome behavior as bridges, early chromosome separation, multiple breaks, and micronuclei release, as well as DNA damage [16]. At low concentrations, however, the nanoparticles often act to promote plant growth and development. Treatment with nanoparticles at low doses enables the detoxification of free radicals and the strengthening of the responses of plants to stress by increasing the activities of antioxidant enzymes [17].

Understanding the molecular mechanisms of environmental abiotic stress responses of plants that were involved for millions of years, is a prerequisite for maintaining global food security. This wide range of molecular mechanisms involves various pathways. However, the process involves a perception of the stress, transduction of stress signals, and regulation of stress-responsive gene expression in transcriptional, translational, posttranslational, and epigenetic levels.

The present chapter summarizes the current understanding of this perception and the transduction mechanisms against the common stressors and emphasizes the stress-specific mechanisms as well as the crossroads in pathways of multiple stressors.

The topic also extended through the emerging environmental stress factors of the twenty-first century, such as microplastics and nanoparticles.

2. Abiotic stresses in plants

2.1 Perception of stress

Plants sense the abiotic stress factors through the physical and chemical alterations, which occur on several biomolecules. Different stress factors may trigger the same or similar stress perception. For instance, drought and salinity have osmotic effects on plant cells and are sensed through similar mechanisms. Under the osmotic stress factors, plant cells are subjected to decreased turgor pressure. The cell membrane sense the reduced tension and allow the transport of the Ca^{2+} into the cell through ion channels. Mechanosensor (MS) ion channels are well preserved in almost all the kingdoms of life. However, it is possible to categorize two distinct mechanosensitive ion channel types. The first type involves microtubules of the cytoskeleton as MSs. For instance, NompC, a member of non-voltage-gated cation channel type transient receptor potential (TRP) ion channels, has helical springs attached to cytoskeleton microtubules and releases currents when microtubules are disturbed [18, 19]. The second type of MSs directly senses the tension on the lipid bilayer of the membrane, such as MscL, TRAAK channel, TREK1 channel, and Piezo1. These channels are found in nearly all animals and sense the osmotic shock, light touch, sound waves, vascular blood flow, and such. In plants, there are five MS ion channel families named as mechanosensitive ion channels of small conductance (MscS)-like (MSL) proteins, two-pore K^+-channels (TPK) and Mid1-Complementing Activity (MCA), reduced hyperosmolality-induced $[Ca^{2+}]_i$ increase (OSCA) channels and Piezo channels [20–24].

MCA channels are different from the other MSs, structurally. MS channels are constituted from multiple transmembrane segments and form multimers. For instance, TPKs form dimers, Piezo channels form trimers, OSCA channels form pentamers and MSL proteins form heptamers. In contrast, the channel formation of MCA is constituted by assembled homotetramer and has only one transmembrane segment. MCA proteins, which are exclusive to plants, are involved in sensing gravity, hardness of soil, and hypo- and hyperosmotic conditions caused by abiotic stress factors. Homologs of other MSs are found in other kingdoms. MSLs are found in archaea, bacteria, protists, and fungi, while Piezo, OSCA, and TPKs are found in animals [22].

MscS directly responds to the membrane tension in bacteria. Homologs of MscS are identified in various subcellular parts of the plant cells. These homologs were named as MscS-like proteins (MSLs). MSL1 is found in the inner membrane of mitochondria. It is ubiquitously expressed and presents a slight anion preference. MSL1 maintains the energy production of mitochondria during osmotic stress by releasing excessive membrane potential and balancing redox homeostasis. MSL2 and MSL3 are found in the inner membrane of chloroplasts. They play roles in maintaining the plastid shape, size, and division under unstressed conditions. The rest of the plant MSLs are found in plasma and endoplasmic reticulum (ER) membranes. They are involved in various physiological processes. For instance, MSL8 is particularly expressed in pollens to take part in maintaining integrity through rehydration, while MSL10 is involved in the jasmonic acid synthesis and signaling during wounding [20].

Two-pore K^+-channels (TPKs) are located on the tonoplast, plasma membrane, and thylakoid membrane. They are crucial for potassium transport from vacuoles to cytosol as the K is the most abundant cation in plants playing important roles in physiological processes (providing turgor, balancing negative charges, enzymatic processes, and protein translation). Members of different K^+ channels can be voltage-gated or, as in TPKs, voltage-independent. Also, in some species as *Arabidopsis* K^+-inward rectifier (Kir)-like channel (KCO_3), they may have one pore only. Since they are insensitive to membrane voltage their activity is dependent on cytoplasmic Ca^{2+} levels, hence they include one or two Ca^{2+} binding EF-hands in the C-terminus. These EF-hands are helix-loop-helix motifs largely found in calcium-binding proteins. In N-terminus, they may contain domains to bind 14–3-3 proteins, which can bind to signaling proteins, such as kinases, phosphatases, and transmembrane receptors. pH value of cytoplasm, trans-tonoplast osmotic gradients, or tension upon cell membrane can lead to TPK channel activity and intracellular osmosensing [21].

In bacteria and animals, cytosolic free Ca^{2+} concentrations $[Ca^{2+}]_i$ tend to increase in response to numerous stimuli as hyperosmolality. However, hyperosmolality-induced $[Ca^{2+}]_i$ increases were widely speculated for plants. Today, reduced hyperosmolality-induced $[Ca^{2+}]_i$ increase (OSCA) channels are known osmosensors, thanks to the calcium-imaging-based forward genetics screens, especially in *Arabidopsis*. They are identified as the previously unknown plasma membrane proteins, which form hyperosmolality-gated calcium-permeable channels in guard cells and root cells. They regulate water transpiration and root growth under stressful conditions. Decreased tension on the lipid bilayer leads OSCA channels to open for Ca^{2+} uptake into the cells [25].

Besides the sensing of lateral tension on the cell membrane, osmotic stresses such as salinity can be sensed through the polarization of the cell membrane. MOCA1 gene-encoded glucurosyl transferase adds glucuronic acid (GlcA) to the plasma membrane. Negatively charged GlcA added glycosil inositol phophoryl ceramides (GIPC) bind Na^+ cations in the membrane and lead to polarization. Salinity-dependent polarization of the membrane results in intracellular Ca^{2+} spikes in plants [26]. High salinity can also be sensed through organizational cell wall alterations. In unstressed conditions, Ca^{2+} assembles the pectin units of the cell wall by cross-linking. Receptor-like kinases (RLKs), such as the FERONIA (FER)-related malectin-domain-containing THESEUS1 and ANXUR1/2 physically interact with pectins in the cell wall. Leucine-rich repeat (LRR) extensins (LRXs) interact with rapid alkalization factor (RALF) peptide ligands and prevent their interaction with FER. Increased salinity, dissociate RALFs from LRXs. RALFs bind to FER and prevent FER to cross-link pectins. Therefore, LRX, RALF, and FER interaction involve in cell wall sensing of salt stress [27].

In this regard, microplastic particle incorporations, such as polystyrene, polypropylene, polyethylene terephthalate, polyethersulfone (PES), or high-density polyethylene (HDPE), are known to decrease soil bulk density after exceeding the 2% (w:w) concentration. If larger microfibers incorporate over even lower amount as 0.2% (w:w), they reduce soil aggregate stability. Moreover, another well-known fact is that plastic films desiccate soil more rapidly in about 0.5–1% concentrations [28]. These mechanical changes can be sensed by plants as reduced tension through the root cap, which is a protective barrier for root meristem. Mechanical stresses and the chemicals that cause toxicity to plants lead root cap cells to excrete mucilage and some other exudates by initiating stress signal transduction pathways. This first line of defense also traps some positively and negatively charged nanoparticles as well as dissolving nutrients and other aggregated soil particles. The more comprehensive effects are discussed in Section 3.

2.2 Transduction of the stress signals

In plants, stress signals lead to stress-related physiological responses through gene expression regulation, epigenetic modifications, small RNA regulation, and more. In cascade of actions, perceived stress requires transduction to trigger specified responses. Signal transduction involves secondary messenger ions and molecules such as Ca^{2+}, abscisic acid (ABA), reactive oxygen species (ROS), and several phospholipids. Some alterations on particular molecules can also transduce the signal as phosphorylation, dephosphorylation, oxidation, nitrosylation, sumoylation, and ubiquitylation in processes called posttranslational modifications (PTMs). Seconder messengers, such as ABA, ROS, and Ca^{2+}, are the most focused signal transducers as they also play different roles in plants. PTMs may alter the regulation, stability, localization, and activity of various cellular components. Therefore, they also play crucial roles in signal transduction.

Abiotic stress responses are complex traits. Phytohormones play fundamental roles in stress adaptation. The most notorious, abscisic acid (ABA) is a small lipophilic sesquiterpenoid (C_{15}) that plays numerous roles in plant growth and development as well as stress response. ABA increases under harsh environmental conditions and serves as a signal transducer for the plant cells to adapt as necessary. Since the discovery of ABA in 1960s, ABA receptors and sensing in plants brought endless debates and controversy. Today, the ABA signaling pathway from a signaling cascade consisting of the PYL ABA receptors, to type 2C protein phosphatases (PP2Cs) and Snf1-related protein kinases 2 (SnRK2s) has been enlightened by multiple structural studies which provided insight on regulating each level of the ABA signal transduction pathway [29]. In this core ABA signaling pathway model, PYLs [14 members; Pyrabactin Resistance 1 (PYR1) and PYR1-like 1–13 (PYL1-PYL13) or regulatory component of ABA Receptor (RCAR1-RCAR14)] bind to ABA to inhibit PP2Cs, which represent a major phosphatase family in plants and consisted of single subunit Mg^{2+}/Mn^{2+}-dependent Ser/Thr phosphatases [30]. ABA response is kept silent at the basal level by negative regulators of ABA as PP2Cs, including ABI1, ABI2, and HAB1. Sucrose non-fermenting 1-related protein kinases (SnRKs) are positive mediators of ABA signaling, which are divided into three groups taking part in metabolic regulation (SnRK1) and abiotic stress signaling (SnRK2 and SnRK3). PP2Cs suppress ABA response by inhibition of positive mediators, especially SnRK2.2, SnRK2.3, and SnRK2.6 [31]. Under stress conditions, elevated levels of ABA induce PYLs to bind and inhibit PP2Cs, which leads to PP2C-dependent SnRK2 inhibition that allows kinases to autophosphorylate and activate effectors as ion channels and ABF transcription factors to activate ABA responses. ABA-responsive elements (ABREs), which generally include PyACGTGG/TC consensus sequence belonging to the G-box family (CACGTG), are *cis*-acting DNA elements with a wide range of gene expression roles. A subgroup of bZIP transcription factors was isolated by using ABRE-binding factors (ABFs). To date, all these AREB/ABF genes have been functionally characterized in *Arabidopsis*. These four genes (ABF1, ABF2/AREB1, ABF3, and ABF4/AREB2) are primarily expressed in vegetative tissues and, the induced ABF1 expression alterations in response to abiotic stress factors are minimal, while ABF2/AREB1, ABF3, and ABF4/AREB2 are significantly up-regulated under ABA and osmotic stresses [32]. Under various abiotic stress conditions, ABA alters 5–10% of plant transcriptome in which half of the alterations are characteristic of drought and salinity responses. ABA-inducible genes include members of Late Embryogenesis Abundant (LEA) proteins called dehydrins, which act as chaperones and protect membrane stability, ROS detoxifying enzymes, and regulatory proteins such as TFs, phosphatases, and kinases. On the other hand, ABA-repressed genes are generally associated with cellular growth [29].

Ca^{2+}, which has limited functions on prokaryotes, is a ubiquitous seconder messenger in plants. Ca^{2+}, which is kept at low concentrations in the cytosol through the calcium pump activities and stored in various intra- and extracellular compartments, influx into the cytosol and presents concentration spikes. These spikes are transduced to cell type-specific and stress-specific signatures depending on timing, intensity, and frequency. For instance, cold shock triggers rapid and transient mainly external Ca^{2+} peaks for seconds, while slow cooling causes two-peaked bimodal elevation for minutes either external or internal sourced as vacuoles. Hyperosmotic stresses cause single or biphasic external or internal Ca^{2+} elevations for a minute, while hypoosmotic stresses lead to rapid bimodal elevations longer than just a minute. Mechanical stresses trigger rapid and transient internal peaks for seconds in contrary to oxidative stresses, which lead to a single external or internal sourced peak for minutes. Anoxia leads to mitochondrial rapid and sustained Ca^{2+}, elevation for hours; however, heat shock causes sustained external or internal increase between 15 and 30 minutes. Alterations in the Ca^{2+} concentration inform targeted cells and organs, subsequently. In plants, calcium sensors are categorized into two groups. The first group, including calmodulins (CaMs) and calcineurin B-like (CBLs) proteins, is called sensor relays. The second group, including calcium-dependent protein kinases (CDPKs), calcium- and calmodulin-dependent protein kinases (CCaMKs), is called sensor protein kinases. The first group which transmits the calcium-induced modification to target proteins has no intrinsic activity. On the other hand, the second members are directly activated upon calcium binding [33]. Calmodulin is a small protein consisting of two pairs EF-hands that bind to Ca^{2+}. CaM globular structure is modified into an open following the conformation Ca^{2+}-binding, which subsequently alters (induces or inhibits) target activity. Therefore, Ca^{2+} signatures are decoded into biochemical responses. There are several genes encoding CaMs and CaM-like proteins (CMLs) in plants diverging by specific sequences or domains to undertake additional properties. For instance, 6 CaM (SpCaM) and 45 CaM-like (SpCML) genes are present for *Solanum pennellii* [34]. Hence, various factors, such as expression pattern alterations, target specificity, affinity to calcium, subcellular localization, or methylation may alter the dynamics of CaM-mediated stress responses. However, CaMs have no active enzymatic activities. They play roles in stress responses over CaM-binding proteins (CaMBPs), which can be further divided into two categories as transduction proteins, such as protein kinases (CBKs) and transcription factors (CBTs). They also interact with effector proteins as ion transporters and enzymes involved in physiological responses. Similar to the CaMs, CBLs are small proteins consisting of two globular domains, which have two EF-hand motifs and connected with a short linker. In contrast to the CaMs, these domains have less conserved variations, resulting in different affinity properties against Ca^{2+} capacities and response specificity [35]. As members of the sensor protein kinases group members, CDPKs contain a protein kinase domain linked to a CaM-like domain by a junction sequence. This sequence is required for kinase activation, which occurs depending on the intramolecular interaction between the CaM-like domain and altered conformation induced by Ca^{2+} binding. As in CaMs, CDPKs are multigenic and encoded by a different number of members in various plant species. Depending on the sensed Ca^{2+} signature, multigenic properties result in variations in activation thresholds, substrate recognition, expression patterns, and sub-cellular localization. CDPKs are located in numerous cellular compartments, such as nucleus, cytosol, chloroplast, peroxisome, ER, and plasma membrane. CBL-interacting protein kinases (CIPKs) complexes are studied extensively. For instance, SOS3 as an EF-hand Ca^{2+} binding protein triggers SnRK3 family kinase SOS2 (CIPK) in SOS (salt overly sensitive) pathway. CDPKs coordinate Ca^{2+} sensing by co-targeting the kinase and its calcium regulator and contribute to cellular response. The other sensor protein kinases known as CCaMKs contain an N-terminal kinase domain and two regulatory domains as CaM-binding domain and 3 EF-hands, which leads to a complex regulatory mechanism involving both Ca^{2+} and Ca^{2+}/CaM binding. Ca^{2+} binding trigger autophosphorylation and elevates CaM affinity. This process removes autoinhibition and leads to kinase activity.

Similar to the Ca^{2+}, variations in ROS species, production source, and accumulation lead to decodable oxidative footprints. During numerous abiotic stress conditions, ROS production presents common oxidative bursts. Common ROS increases are balanced through detoxification and scavenging. On the other hand, ROS interacts with signaling molecules and transduces them to specific stress-related responses. Most of the ROS have a local site of action due to the short half-life and inadequate stability to diffuse to long range, except H_2O_2. Hence, they are not capable of crossing cellular membranes [36]. However, ROS signaling can be induced in either extrinsic or intrinsic manner. An extrinsic path involving apoplast and cell wall facilitates respiratory burst oxidase homologs (RBOHs), aquaporins (AQPs), and cell wall-bound peroxidases (PRXs) for signaling. RBOHs generate superoxide radicals at the apoplast by using cytosolic NADPHs. RBOH activity is initiated by EF-hand domains following the Ca^{2+} binding. They can be either activated or inactivated in response to various stress/stimuli to trigger ROS signatures at the apoplast. As well as cell wall-bound peroxidases other oxidases on apoplast can produce ROS as well. RBOHs can also be activated through phosphorylation, biquitylation, persulfidation, nitrosylation, and glutathionylation. Apoplastic ROS infiltrates the cell through AQPs, which are regulated by phosphorylation, acetylation, and/or guanidinylation. Following these posttranslational modifications elevated ROS in cytosol triggers cytosolic phosphorylation reactions through receptors. Therefore, the extrinsic ROS pathway plays an important regulatory role by altering the influx of Ca^{2+} through the channels. As described earlier, the intrinsic path involving cytosol and nucleus has numerous signaling components, such as MAPK cascades, CDPKs, CIPKs, and phosphatases, such as PP2A, PP2C, and PTPs, and signaling molecules, such as Ca^{2+} and various phytohormones. Cytosolic H_2O_2 levels are balanced through the activity of AQPs. They are located on both organelle and cell membranes and can transport H_2O_2 both ways. Sensor relay proteins may lead to ROS scavenging and therefore alter signaling and the related stress responses. Cytosol is a central hub to decode and transduce ROS signatures leading to specific responses. Therefore, the majority of the stress-related TFs have either ROS- or redox-dependent activation to initiate transcription in the nucleus. Moreover, inside the plant cell, organelles participate in ROS regulation by both production and scavenging. Organelle-derived ROS communicates through different organelles and nucleus. Organelle-autonomous regulation, nucleus-controlled retrograde/anterograde regulation, and direct export/import are deciding factors for ROS accumulation. Organelle-derived ROS communication is short-ranged. They either cross the cytosol at a very limited distance or do not cross at all [37].

As described earlier, each abiotic and biotic stresses transduce cell type-specific and stress-specific signatures, depending on timing, intensity, and frequency. Microplastics as polystyrene significantly elevates MAPK signaling pathway similar to plant—pathogen interaction. It also upregulates metabolic processes such as microtubule-based movement and cell cycle as well as down-regulating response to reactive oxygen species and oxidative stress. Through transducing the signals, microplastics affect the synthesis of a variety of amino acids, such as alanine, aspartate, and glutamate, which are speculated to be highly related to the mechanism of plant tolerance or detoxification. Plants enhance a variety of metabolic pathways to increase resistance by producing various metabolites. Polystyrene alters plant hormone biosynthesis, phenylpropanoid biosynthesis, and fatty acid metabolism pathways significantly, which take part in important stress signal transduction pathways and negatively affect transcriptional levels [38].

3. Micro/nanoplastics and nanoparticles

In contrast to the well-studied molecular mechanisms of abiotic stresses described earlier, a limited number of investigations have been conducted to understand plant responses to micro/nanoplastics. Nevertheless, the effects of microplastics and nanoplastics on plant germination, morphology, physiology, and plant responses are substantial.

Exposure to microplastics and nanoplastics in the ecosystem has often been reported to have adverse effects on many organisms and the environment. Negative impact on the activity of soil microorganisms and animal species [39, 40]; damage to soil structure by altering the physical and chemical properties of soil [41]; adverse effect on plant growth [14, 39], decreased growth and productivity of aquatic flora [42], and negative impact on nitrogen and carbon cycles [43] are some major of these effects. Microplastic pollution affects the nitrogen (N) cycle in soil ecosystems by significantly reducing NH_3 volatilization. Microplastics change dissolved organic material fractions and soil properties [44].

However, some researchers have reported that exposure to microplastics does not cause plant, soil animal, or soil damage and may even play a role in preparing a more favorable environment for the growth of plants and soil animals [45]. Current soil microplastic research has shown that agricultural soils are subject to higher environmental exposure than other environments [39]. Agricultural soil has been recognized as a major pollutant sink for microplastics and nanoplastics that can affect ecosystem and biodiversity. Microplastics are responsible for many changes in the physicochemical properties of the soil, such as soil nutrient balance, soil porosity, fertility, aggregate stability, bulk density, enzyme activity, and water-holding capacity [14]. Many of these changes affect the soil microbiome as well as terrestrial plants. Low-density polyethylene (LDPE) and bio microplastics trigger significant changes in rhizosphere bacterial communities. These changes have proven to have far-reaching effects on soil nutrient cycling and plant health in agroecosystems [46]. Findings on the effects of microplastics, especially on the development of agricultural products and soil health, are very important in the development of policies related to agricultural sustainable development.

Especially in recent years, biodegradable plastics such as mulch films used in agricultural lands, organic fertilizer and sewage sludge application, greenhouse materials, soil conditioners for agriculture, irrigation with reclaimed water, and biosolids pose a significant threat to the functioning of the agroecosystem [47]. Mulch films can degrade soil quality as well as reduce crop production [48]. Studies using a global meta-analysis based on field/laboratory measurement have shown that undiminished accumulation of microplastics adversely affects crop growth and soil health [47]. It has been reported that the effects of microplastics on plants are affected by the microplastic properties and concentration, such as particle type, size, shape, polymer structure, and vary among plant species [49–51].

The inhibitory mechanisms of microplastics are divided into two direct and indirect mechanisms. Direct mechanisms include blocking pores or light, causing mechanical damage to root, inhibiting gene expression, and releasing additives. On the other hand, indirect mechanisms include changes in soil properties and the influencing of soil microorganisms and animals [52]. The presence of microplastics in the soil causes a decrease in nutrient availability and microorganism activity, which ultimately affects the growth, development, and production of crops [49, 53, 54]. It has been shown that microplastics can inhibit the growth and performance of higher plants as well. In addition, leachate from coastal microplastics has been reported to reduce seed germination and early plant growth. Coastal ecosystems are threatened by exposure to coastal microplastics [55].

Microplastics can inhibit germination and seedling development, delay fruit ripening, and reduce yield. It also causes a decrease in biomass and modulates the growth indices of the plant. Physiological responses of plants to microplastics include disruption of cellular homeostasis, induction of oxidative stress, changes in antioxidative enzyme activities, and photosynthetic parameters [13]. Microplastics and nanoplastics cause oxidative stress by being transported from root to stem on terrestrial plants or by being directly absorbed by leaves and accumulating in various tissues, and adversely affect plant growth parameters by altering metabolism, photosynthesis processes, and related gene expression levels. Once microplastics and nanoplastics enter the plant, they cause different effects on roots, including impairing water and nutrient uptake and reducing transpiration rate [56]. When the effect of microplastics polyethylene on photosynthetic performance in the leaves of *Nicotiana tabacum* seedlings was evaluated by transcriptomic analysis, 79 DEGs related to photosynthetic proteins were detected. It was reported that most of the genes were downregulated under high microplastic concentrations. The reduction in photosynthetic capacity is due to the expression of genes involved in light collecting, electron transport, and photosystem function in chloroplasts. Modulation of photosynthetic capacity includes high ROS accumulation, inhibition of leaf pigment synthesis and Rubisco activity, reduction of light utilization and dark respiration, and inhibition of electron transport between PSII and PSI [57].

Identification and characterization of microplastics in plant tissues have been achieved through Raman confocal microscopy and mass spectrometry-based approaches (Py-GC–MS and ICP-MS) that map and characterize cross-sections of plant tissue [56]. Fluorescent and confocal microscopy studies have shown that microplastics physically block the pores in the seed capsule, delay germination, and also affect root development due to their small size and high adsorption capacity [58]. Nanoplastics can accumulate in plants at different levels, depending on their surface charge. Positively charged nanoplastics accumulate at relatively low levels at root tips but result in higher reactive oxygen species accumulation. They inhibit plant growth and seedling development more strongly than negatively charged nanoplastics. In contrast, negatively charged nano plastics are frequently observed in the apoplast and xylem [59]. It is also reported that polystyrene nanoplastics (22–24 nm) accumulate in plants, depending on their surface charge. Positively charged polystyrene nanoplastics accumulate more readily than negatively charged polystyrene nanoplastics. The photosynthesis inhibition efficiency of positively charged nanoplastics is higher and the activation of the antioxidant system can be stimulated more strongly [60]. Microplastics show different effects at different concentrations. Concentration-dependent developmental alterations are summarized in **Table 1**. On the other hand, it has also been shown that PE in the soil has either non or slight significance on plant growth [89, 90]. The same is applied to microplastics [90]. Nevertheless, microplastics at high concentrations also may not have a significant effect on plant growth, photosynthesis, and reactive oxygen species (ROS) content depending on the particle size [91].

Although the uptake mechanisms of plastic particles cannot be clearly explained, it can be achieved through numerous entry routes, such as passive diffusion through the cell wall, endocytosis, facilitated diffusion through carrier proteins of the cell membrane, stomatal opening, and intercellular translocation mechanisms *via* plasmodesmata [14]. In particular, microplastics that are effectively taken from the lateral root regions by the crack-entry mode are then transported from the roots to the shoots *via* the xylem [92]. Generally, for nanoplastics and occasionally larger microplastics, translocation can occur from leaf to root *via* stomata or from root to leaf *via* apoplastic transport [51]. Micro- and nanoplastics on the root surface and root hairs can change the shape of root epidermal cells and block the cell junctions and cell wall pores, thereby preventing root hairs from absorbing water and nutrients, resulting in reduced plant growth [59]. Micro- and nanoplastics have been found in various plant tissues and organs, such as root and stem xylem, leaf [45], stem [92], and root [93, 94]. Microplastics retained by plant roots can become part of the plant body and eventually join the higher food chain and be stored in the edible parts of plants [94]. Microplastics also play a role in the transport of various toxic chemicals, such as polycyclic aromatic hydrocarbons (PAHs), potentially toxic elements (PTEs), and antibiotics. Microplastics and nanoplastics can act as a carrier for heavy metals derived from the environment, leading to higher heavy metal accumulation in plant leaves [95]. The accumulation of micro- and nanoplastics and subsequent damage to plants further affect crop productivity, food safety, and quality and lead to potential health risks.

Studies in many plant species show that nanomaterials can be absorbed by the roots and transported to other organs. However, the amount of absorbance varies greatly by the physicochemical properties of nanomaterials, the plant species, and environmental conditions [15]. The biological effects of nanoparticles also vary greatly depending on their physicochemical properties, plant tissue, the environment they encounter with the plant, surface charge, size, and concentration. Nanoparticles enter plant tissues through pores in the cell wall, stomata, and crack enter. Size is the most important parameter for absorption into plant tissues. The negative charge of the cell wall allows positively charged particles to penetrate the surface more easily [96]. The number of polar and nonpolar groups in the structure of the material, in other words, its hydrophobicity, is an important determining.

To obtain nanomaterials, two basic methods are used as top-down and bottom-up production. In the method called top-down, the whole material is processed and divided into small pieces. Milling, etching, electro-explosion, sputtering, laser ablation, lithography, aerosol-based techniques, and liquid-phase techniques are frequently used in top-down. In the bottom-up production method, the material is obtained by synthesizing atoms and molecules through chemical reactions. Chemical vapor deposition (CVD), chemical vapor condensation (CVC), molecular beam epitaxy (MBE), plasma arcing, and wet chemical methods are used for bottom-up.

Plant	Microplastics-Nanoplastics	Plant Response	Concentration	Ref
Arabidopsis	Poly (butylene adipate-co-terephthalate)	Disrupted the photosynthetic system	20 g kg^{-1}	[48]
Arabidopsis	PE, PET, PVC, PVC, PS	Changes in phenotypic, metabolic and transcriptional profiles	1 mg mL^{-1}	[61]
Brassica oleracea	PE	Changes in enzymatic factors, modifications in antioxidant defense system	0.01, 0.1, 1, 10, 100, 1000, and 10,000 mg L^{-1}	[62]
Centaurea cyanus	PVC	Reduced plant growth and photosynthetic efficiency	1%, w/w	[63]
Cucumis sativus	PE-MS	Changes in metabolic profile	10, 100, and 1000 mg L^{-1}	[64]
Cymodocea nodosa	PS	Lower numbers of leaves per shoot, reduced photochemical efficiency, and increased pigment content	68 µg/L	[65]
Fragaria x *ananassa*	HDPE	Decreased plant height, lower yield	0.2 g kg^{-1}	[66]
Glycine max	Polylactic acid microplastics	Decreased the root length, Changes in amino acid metabolism	0.1% w/w	[67]
Hordeum vulgare	PE-MS	Changes in metabolic profile	10, 100, and 1000 mg L^{-1}	[64]
Hydrilla verticillate	Polystyrene nanoplastics and bisphenol F	Decreased relative growth rate and chlorophyll content, triggered antioxidant responses	10 mg L^{-1}	[68]
Lactuca sativa	Differentially charged PS	Growth inhibition, root lignification, root cell apoptosis, oxidative stress responses, accelerated chlorophyll decomposition and hampered normal electron transfer	30 mg L^{-1}	[69]
L. sativa	PEF, fossil-based plastic PET	Inhibited growth, photosynthesis, and the accumulation of other nutrients	0.5%, 1.0%, and 2.0% w/w	[70]
Lemma minor	PE	Inhibited growth rate and chlorophyll content	50 mg L^{-1}	[71]
Lemna minor	PE	Tolerated the presence of MPs for a long period of time	100 mg L^{-1}	[72]
Lens culinaris	PE	Reduced germination viability and plant growth	10, 50, and 100 mg L^{-1}	[73]
Lycopersicon esculentum	PS, PE, and PP	Adverse effects on seed germination, root growth, and physiological and biochemical activities	10, 100, 500, and 1000 mg L^{-1}	[74]
Oryza sativa	PE and biodegradable mulch films	Reduced the height and dry weight, induced oxidative stress, changes in transcriptional profile	1% w/w	[75]
O. sativa	PS and Phe	Inhibited growth, improved antioxidant potential, destroyed the photosynthetic system	50 mg L^{-1} (PS), 1 mg L^{-1} (Phe)	[76]

Plant	Microplastics-Nanoplastics	Plant Response	Concentration	Ref.
O. sativa	PE	Reduced growth parameters, changes in biomass accumulation, physiological and biochemical attributes	250 mg and 500 mg L^{-1}	[77]
O. sativa	PS and PVC	Reduced plant growth and photosynthetic rate	1.5 and 3.0 mg L^{-1}	[78]
Pistia stratiotes	PS	No effect on plant growth negatively affected the translocation of Bisphenol S	10 mg kg^{-1}	[79]
Raphanus sativus	PVC	Reduced plant growth	2%, w/w	[80]
R. sativus	PE	Changes in enzymatic factors, modifications in antioxidant defense system	0.01, 0.1, 1, 10, 100, 1000, and 10,000 mg L^{-1}	[62]
Senecio inaequidens	PVC	Reduced plant growth and photosynthetic efficiency	1%, w/w	[63]
Solanum lycopersicum	MFB and MFL	Concentration-dependent decline in growth	0.4, 2.4, 4.4, 6.4, and 8.4% (w/w)	[81]
S. lycopersicum	PE-MS	Changes in metabolic profile	10, 100, and 1000 mg L^{-1}	[64]
Solanum nigrum	LDPE	Inhibited growth index for medium-high concentration	0.135, 0.27, 0.81, and 1.35 mg kg^{-1}	[82]
Thinopyrum junceum	HDPE and PP	Reduced plant growth	0.0125–0.1 mg/ml	[55]
Trigonella foenum -graecum	LDPE	Increased plant height	1 g kg^{-1}	[83]
Triticum aestivum	PS and DMF	Reduced plant height and base diameter of seedlings	10 and 100 mg kg^{-1} 1% DMF	[84]
Vigna radiata	Shoe sole fragments	Adverse effects on plants	size: 57–229 μm	[85]
Zea mays	PHBV	Reduced plant growth and foliar nitrogen	0.01%, 0.1%, 1%, and 10%	[86]
Z. mays	PMF	Limited crop growth and N uptake	0.5% w/w	[87]

PE: Polyethylene, PET: Polyethylene terephthalate, PVC: Polyvinylchloride, PS: polystyrene, PE-MS: Polyethylene microspheres, HDPE: High-density polyethylene, PEF: Polyethylene 2,5-furan-dicarboxylate, PP: Polypropylene, Phe: Phenanthrene, MFB: Microfibers, MFL: Microfilms, LDPE: Low-density polyethylene, PP: Polypropylene DMF: Degradable mulching film, PHBV: Bioplastic poly (3-hydroxybutyrate-co-3-hydroxyvalerate, PMF: Polyester microplastic fibers factor as well as the size and particle charge. Hydrophobicity can be an important limiting factor not only in the uptake from the cell wall but also in its transport within the organism [88].

Table 1.
Concentration-dependent developmental alterations of micro- and nanoplastics on plants.

Atoms and molecules are brought together in a controlled manner to form larger systems, clusters, organic lattices, multi-molecular structures, and synthesized macro-molecules [97]. Nanomaterials can be divided into four classes according to their structures as: (i) zero-dimensional nanomaterials with all dimensions on the nanoscale, (ii) one-dimensional nanomaterials with two dimensions at the nanoscale, (iii) two-dimensional nanomaterials with one dimension at the nanoscale, (iv) three-dimensional nanomaterials, all of which are microscale. These materials may have organic (carbon-based) or inorganic content [98].

Nanomaterials show better optical, electrical, and magnetic properties due to their high reactivity and mechanical resistance. These materials, which have completely new features unlike the materials we know, have gained great popularity in the fields of health, chemistry, cosmetics, food, military practices, and agriculture. However, the great advantages of these materials also brought new risks and uncertainties.

The nanoparticles can penetrate leaf tissue through stomata or cuticles. The cuticle layer is the outermost barrier of the leaf. It may restrict the entry of nanoparticles smaller than 5 nm. Movement of nanoparticles between 10 and 50 nm occurs through the adjacent cell's cytoplasm referred as a symplastic route. Therefore, larger nanoparticles between 50 and 200 nm are translocated between the cells known as apoplastic route. Following the penetration, nanoparticles interrupt the electron transport chain (ETC) cycle of chloroplast and mitochondria and trigger oxidative bursts. The excess level of ROS leads plant to destruction of DNA, oxidation of proteins, peroxidation membrane lipids and ultimately programmed cell death (PCD) [99].

With nanomaterials, it is tried to develop strategies to increase food quality, protect against pests and diseases, determine species, make instant interventions by monitoring the development of the plant, and increase yield [100]. It is used extensively in the form of nano fertilizers, nano herbicides, nano fungicides, and nano pesticides in order to ensure the controlled release of necessary chemicals, especially without disturbing the environmental ecosystem and to reduce the number of pesticides and herbicides dispersed into the environment [101]. Nanoparticles are also frequently used in agriculture to improve soil quality, increase germination, support plant growth, and increase yield. In addition, its use for reducing the effects of abiotic stresses and reversing the damage caused is a subject of intense research. There is also a lot of research on gene transfer and obtaining new transgenic plants through nanoparticles [102].

Nanomaterials often accumulate in the soil and encounter plants through roots. Depending on their size, the free nanomaterials in the soil penetrate the epidermal cells in the root tissues with the effect of osmotic pressure and capillary forces. The wall of epidermal cells of root tissue acts as a semipermeable barrier with gaps smaller than 20 nm and prevents the passage of large particles. Nanomaterials penetrating through the cavities in the cell wall are transported up to the central cylinder apoplastically through intercellular spaces or symplastically *via* plasmodesmata. Plasmodesmata are about 40 nm wide [103]. For the nanomaterials to enter the phloem and xylem and to be transported to the shoots and other organs, they must pass through the central cylinder. This happens by binding to the endodermal cell membrane's carrier proteins through endocytosis, pore formation, and transport. Otherwise, the nanomaterials accumulate in the casparian strip and cause textural damage. Here, soil content and nanomaterial interaction also appear as limiting factors in the amount of penetration from plant roots [104].

Although nanomaterials are often taken from the soil *via* roots, in some cases, they can also be taken up through leaves or other organs. The cuticle layer on the leaves acts as a barrier for materials larger than 5 nm. Again, stomata on the leaf surface provide passage for materials smaller than 40 nm. Generally, the materials entering the tissue with a size of up to 50 nm are transported by symplastic means, while the materials in the 50–200 nm range are transported apoplastically [105]. Nanomaterials can be transported bidirectionally with sugar and other soluble organic compounds in the phloem [106]. It can be carried to the roots, stems, fruits, and seeds and accumulate there. Leaf morphology, chemical composition of the genus, presence of trichomes, presence of leaf exudates, wax layer, and most importantly size are the most important factors in the penetration of nanomaterials into the tissue by adhering to the leaf surface [107].

In some cases, nanomaterials cause cell wall damage, promoting the formation of new pores. The *de novo* pores formed in the cell wall can simply enter the cell. Similarly, it can enter the cell by imitating the behavior of biological components or by forming a bond with chemicals in the external environment [108]. It can progress through symplastic and apoplastic pathways in the cell, as well as enter and accumulate in organelles with energy-dependent endocytosis. Accumulation at high concentrations causes toxicity to the cell.

Despite the long-standing experimental contribution to the topic, there are still large gaps to be addressed. Some of the key aspects are as follows:

- Long-term studies are required to diversify the future projections about the fate of micro/nanoplastics and nanoparticles in different environments (soil, aquatic systems, etc.) and to understand the uptake potential of micro/nanoplastics and nanoparticles by plants and bioaccumulation of micro/nanoplastics and nanoparticles in plants.

- It is necessary to determine the uptake potential of micro/nanoplastics and nanoparticles by the plant and their positive or negative effects on plant growth, according to their various sizes and surface charges (positively or negatively charged). Moreover, much more data are required to evaluate the toxicity of micro/nanoplastics and nanoparticles in plants.

- Data on the morphological, physiological, and omic-based (genomic, transcriptomic, proteomic, and metabolomic) evaluation of the effects of micro/nanoplastics and nanoparticles on plants in different plant species is quite limited. There is a requirement to increase the number of studies at the molecular level to understand the interactions between micro/nanoplastic and nanoparticles in plants and to illuminate the related molecular pathways.

- Determining the changes that micro/nanoplastics may cause in the physical and chemical structure of the soil and the indirect effects of these plastics on the development of plants will provide extremely beneficial information for agricultural improvement.

- More data are required to better understand the interactions of micro/nanoplastics with soil microorganisms and animals under different soil conditions. In addition, studies to understand the effects of micro/nanoplastics on nitrogen and carbon cycles will be extremely beneficial in terms of the evaluation of environmental stress on plants and novel strategies for agricultural improvement.

- There is very limited data on the interactions between nanoplastics and other nanoparticles. In this regard, precise data are required to elucidate the cumulative effect of nanoparticles on plants.

- Limited data are available regarding the bioaccumulation of nanoplastics and nanoparticles in plant tissues. Increasing these data will be extremely helpful in understanding the fate of them in the food chain.

4. Concluding remarks on crossroads between abiotic stresses and micro/nanoplastics and nanoparticles

Adsorption or uptake of micro/nanoplastics through root tips leads to toxicity and activates mechanoreceptors similar to other osmotic stresses. Apparently, plants share common stress perception and signal transduction crossroads between familiar abiotic factors and novel micro/nanoplastics and nanoparticles, even though evolutionary adaptations have not introduced these pollutants to plants before. Micro/nanoplastics may cause mechanical stress by physical blockage, disconnecting cells, and consequently reducing signal transmission. Obstructing properties prevent plants to uptake water and nutrients, hence reducing germination. Due to the extremely small size of nanoparticles, rapid and relatively uncontrolled penetration and translocation to various cell compartments occur. Transport proteins or ion channels mentioned earlier utilize the proportional entry of nanoparticles as well as endocytosis. Especially metallic forms of nanoparticles, such as Cu, Ni, Zn, TiO_2, and CeO_2, may lead to excessive ROS production through the Fenton reactions by altering oxidative states. Occasionally, nanoparticles can decrease intracellular H_2O_2 concentrations and lipid peroxidation by increasing the efficiency of redox reactions by playing a central role in electron retransmission. On the other hand, nanoparticle exposure may increase the production of $^1O_2$; hence, creating unique ROS signatures to be decoded to appropriate abiotic stress response. Ca^{2+} ions play a vital role in increasing plant tolerance during abiotic stresses by modulating stress signaling and responses. Ca nanoparticle applications lead to better utilization of mineral elements. CaO nanoparticles undertake a key role in stress signaling processes to maintain ion homeostasis in plants. ABA is a notorious phytohormone for stress signaling and abiotic stress responses. Nanoparticle exposure rapidly impacts the ABA signaling pathway. Ag nanoparticles induced ABA signaling by enhancing ROS and altering root growth. La_2O_3 nanoparticles induce rapidly detectable ABA fluctuations through the ABA receptors. Similarly, ZnO nanoparticles exposure mediates the transcript level of ABA synthesis and catabolism-related genes. Further proteomic, transcriptomic, epigenetic, and other omic-based examinations will provide insight into the regulatory role of nano-sized pollutants in the stress resistance of future plant cultivars [109–112].

Author details

Tamer Gümüş, Sinan Meriç, Alp Ayan* and Çimen Atak
Faculty of Science and Letters, Department of Molecular Biology and Genetic, Istanbul Kultur University, Istanbul, Turkey

*Address all correspondence to: a.ayan@iku.edu.tr

References

[1] Zahan M, Gandhi M. Environmental Laws in India - An analysis. In: Sengupta P, Bhattachrya B, editors. Sustainable Development and Environment Governances. Adhyayan Publishers; 2018. pp. 75-88

[2] Morris JL, Puttick MN, Clark JW, Edwards D, Kenrick P, Pressel S, et al. The timescale of early land plant evolution. Proceedings of the National Academy of Sciences of the United States of America. 2018;**115**(10):E2274-E2283. DOI: 10.1073/pnas.1719588115

[3] Çelik Ö, Ayan A, Atak Ç. Enzymatic and non-enzymatic comparison of two different industrial tomato (Solanum lycopersicum) varieties against drought stress. Botanical Studies. 2017;**58**(1):32. DOI: 10.1186/s40529-017-0186-6

[4] de Vries J, Archibald JM. Plant evolution: Landmarks on the path to terrestrial life. The New Phytologist. 2018;**217**(4):1428-1434. DOI: 10.1111/nph.14975

[5] Catarino B, Hetherington AJ, Emms DM, Kelly S, Dolan L. The stepwise increase in the number of transcription factor families in the Precambrian predated the diversification of plants on land. Molecular Biology and Evolution. 2016;**33**(11):2815-2819. DOI: 10.1093/molbev/msw155

[6] Maarten JMC, WWB J. The number of known plants species in the world and its annual increase. Phytotaxa. 2016;**261**(3):201-217. DOI: 10.11646/phytotaxa.261.3.1

[7] Shanmugam V, Das O, Rasoul &, Neisiany E, Babu K, Singh S, et al. Polymer recycling in additive manufacturing: An opportunity for the circular economy. Materials Circular Economy. 2020;**2**(1):1-11. DOI: 10.1007/S42824-020-00012-0

[8] Geyer R, Jambeck JR, Law KL. Production, use, and fate of all plastics ever made. Science Advances. 2017;**3**(7):e1700782. DOI: 10.1126/sciadv.1700782

[9] Eriksen M, Lebreton LCM, Carson HS, Thiel M, Moore CJ, Borerro JC, et al. Plastic pollution in the World's oceans: More than 5 trillion plastic pieces weighing over 250,000 tons afloat at sea. PLoS One. 2014;**9**(12):e111913. DOI: 10.1371/journal.pone.0111913

[10] Yousif E, Haddad R. Photodegradation and photostabilization of polymers, especially polystyrene: Review. Springerplus. 2013;**2**(1):1-32. DOI: 10.1186/2193-1801-2-398

[11] Lambert S, Wagner M. Formation of microscopic particles during the degradation of different polymers. Chemosphere. 2016;**161**:510-517. DOI: 10.1016/J.chemosphere.2016.07.042

[12] Duis K, Coors A. Microplastics in the aquatic and terrestrial environment: Sources (with a specific focus on personal care products), fate and effects. Environmental Sciences Europe. 2016;**28**(1):1-25. DOI: 10.1186/S12302-015-0069-Y

[13] Kumari A, Rajput VD, Mandzhieva SS, Rajput S, Minkina T, Kaur R, et al. Microplastic pollution: An emerging threat to terrestrial plants and insights into its remediation strategies. Plants. 2022;**11**(3):340. DOI: 10.3390/plants11030340

[14] Yadav S, Gupta E, Patel A, Srivastava S, Mishra VK, Singh PC, et al. Unravelling the emerging threats of microplastics to agroecosystems. Reviews in Environmental Sciences Bio/Technology. 2022;**21**(3):771-798. DOI: 10.1007/s11157-022-09621-4

[15] Chichiriccò G, Poma A. Penetration and toxicity of nanomaterials in higher plants. Nanomater. 2015;**5**(2):851-873. DOI: 10.3390/nano5020851

[16] Castiglione MR, Giorgetti L, Geri C, Cremonini R. The effects of nano-TiO2 on seed germination, development and mitosis of root tip cells of Vicia narbonensis L. and Zea mays L. Journal of Nanoparticle Research. 2011;**13**(6):2443-2449. DOI: 10.1007/S11051-010-0135-8

[17] Tariq M, Choudhary S, Singh H, Siddiqui MA, Kumar H, Amir A, et al. Role of nanoparticles in abiotic stress. In: Technology in Agriculture. London, UK, London, UK: IntechOpen; 2021. DOI: 10.5772/intechopen.99928

[18] Basu D, Haswell ES. Plant mechanosensitive ion channels: An ocean of possibilities. Current Opinion in Plant Biology. 2017;**40**:43-48. DOI: 10.1016/j.pbi.2017.07.002

[19] Wang X, Niu D, Zhang Y, Jia L, Zhang YJ, et al. The NompC channel regulates Nilaparvata lugens proprioception and gentle-touch response. Insect Biochemistry and Molecular Biology. 2019;**106**:55-63. DOI: 10.1016/j.ibmb.2018.11.005

[20] Li Y, Hu Y, Wang J, Liu X, Zhang W, Sun L. Structural insights into a plant mechanosensitive Ion Channel MSL1. Cell Reports. 2020;**30**(13):4518-4527.e3. DOI: 10.1016/j.celrep.2020.03.026

[21] Dabravolski SA, Isayenkov SV. New insights into plant tpk ion channel evolution. Plants. 2021;**10**(11):2328. DOI: 10.3390/plants10112328/s1

[22] Yoshimura K, Iida K, Iida H. MCAs in Arabidopsis are Ca2+−permeable mechanosensitive channels inherently sensitive to membrane tension. Nature Communications. 2021;**12**(1):1-9. DOI: 10.1038/s41467-021-26363-z

[23] Zhang M, Wang D, Kang Y, Wu JX, Yao F, Pan C, et al. Structure of the mechanosensitive OSCA channels. Nature Structural & Molecular Biology. 2018;**25**(9):850-858. DOI: 10.1038/s41594-018-0117-6

[24] Mousavi SAR, Dubin AE, Zeng WZ, Coombs AM, Do K, Ghadiri DA, et al. PIEZO ion channel is required for root mechanotransduction in Arabidopsis thaliana. Proceedings of the National Academy of Sciences of the United States of America. 2021;**118**(20):e2102188118. DOI: 10.1073/pnas.2102188118

[25] Yuan F, Yang H, Xue Y, Kong D, Ye R, Li C, et al. OSCA1 mediates osmotic-stress-evoked Ca2+ increases vital for osmosensing in Arabidopsis. Nature. 2014;**514**(7522):367-371. DOI: 10.1038/nature13593

[26] Jiang Z, Zhou X, Tao M, Yuan F, Liu L, Wu F, et al. Plant cell-surface GIPC sphingolipids sense salt to trigger Ca2+ influx. Nature;**572**(7769):341-346. DOI: 10.1038/s41586-019-1449-z

[27] Feng W, Kita D, Peaucelle A, Cartwright N, Doan V, et al. The FERONIA receptor kinase maintains cell-wall integrity during salt stress through Ca2+ signaling. Current Biology. **28**(5):666-675. DOI: 10.1016/j.cub.2018.01.023

[28] Joos DTC. Soil under stress: The importance of soil life and how it is influenced by (micro) plastic pollution.

Computational and Structural Biotechnology Journal. 2022;**20**:1554-1566. DOI: 10.1016/j.csbj.2022.03.041

[29] Ng Ley M, Melcher K, Teh B, Xu HE. Abscisic acid perception and signaling: Structural mechanisms and applications. Acta Pharmacologica Sinica. 2014;**35**(5):567-584. DOI: 10.1038/aps.2014.5

[30] Janicki M, Marczak M, Cieśla A, Ludwików A. Identification of novel inhibitors of a plant group a protein phosphatase type 2C using a combined In Silico and biochemical approach. Frontiers in Plant Science. 2020;**11**:1416. DOI: 10.3389/fpls.2020.526460

[31] De Dios B-LJ, Moreno JR, Gamez-Arjona FM, Pardo JM, Punkkinen M, Zhu JK, et al. Upstream kinases of plant Sn RK s are involved in salt stress tolerance. The Plant Journal. **93**(1):107-118. DOI: 10.1111/tpj.13761

[32] Li F, Mei F, Zhang Y, Li S, Kang Z, Mao H. Genome-wide analysis of the AREB/ABF gene lineage in land plants and functional analysis of TaABF3 in Arabidopsis. BMC Plant Biology. 2020;**20**(1):1-15. DOI: 10.1186/s12870-020-02783-9

[33] Xu T, Niu J, Jiang Z. Sensing mechanisms: Calcium signaling mediated abiotic stress in plants. Frontiers in Plant Science. 2022;**13**:2030. DOI: 10.3389/fpls.2022.925863

[34] Shi J, Du X. Identification, characterization and expression analysis of calmodulin and calmodulin-like proteins in Solanum pennellii. Science Reports. 2020;**10**(1):1-17. DOI: 10.1038/s41598-020-64178-y

[35] Zeng H, Xu L, Singh A, Wang H, Du L, Poovaiah BW. Involvement of calmodulin and calmodulin-like proteins in plant responses to abiotic stresses. Frontiers in Plant Science. 2015;**6**:600. DOI: 10.3389/fpls.2015.00600

[36] Dvořák P, Krasylenko Y, Zeiner A, Šamaj J, Takáč T. Signaling toward reactive oxygen species-scavenging enzymes in plants. Frontiers in Plant Science. 2021;**11**:618835. DOI: 10.3389/fpls.2020.618835

[37] Huang H, Ullah F, Zhou DX, Yi M, Zhao Y. Mechanisms of ROS regulation of plant development and stress responses. Frontiers in Plant Science. 2019;**10**:800. DOI: 10.3389/fpls.2019.00800

[38] Lu S, Chen J, Wang J, Wu D, Bian H, Jiang S, et al. Toxicological effects and transcriptome mechanisms of rice (*Oryza sativa* L.) under stress of quinclorac and polystyrene nanoplastics. Ecotoxicology and Environmental Safety. 2023;**1**(249):114380-114392. DOI: 10.1016/j.ecoenv.2022.114380

[39] Qiu Y, Zhou S, Zhang C, Zhou Y, Qin W. Soil microplastic characteristics and the effects on soil properties and biota: A systematic review and meta-analysis. Environmental Pollution. 2022;**313**:120183. DOI: 10.1016/j.envpol.2022.120183

[40] Ji Z, Huang Y, Feng Y, Johansen A, Xue J, Tremblay LA, et al. Effects of pristine microplastics and nanoplastics on soil invertebrates: A systematic review and meta-analysis of available data. Science of the Total Environment. 2021;**788**:147784. DOI: 10.1016/j.scitotenv.2021.147784

[41] Qi Y, Beriot N, Gort G, Huerta Lwanga E, Gooren H, Yang X, et al. Impact of plastic mulch film debris on soil physicochemical and hydrological properties. Environmental Pollution.

2020;**266**:115097. DOI: 10.1016/j.envpol.2020.115097

[42] Ghosh D, Sarkar A, Basu AG, Roy S. Effect of plastic pollution on freshwater flora: A meta-analysis approach to elucidate the factors influencing plant growth and biochemical markers. Water Research. 2022;**225**:119114-119127. DOI: 10.1016/j.watres.2022.119114

[43] Shen M, Song B, Zhou C, Almatrafi E, Hu T, Zeng G, et al. Recent advances in impacts of microplastics on nitrogen cycling in the environment: A review. Science of the Total Environment. 2022;**815**:152740. DOI: 10.1016/j.scitotenv.2021.152740

[44] Feng Y, Chen S, Han L, Wang B, Sun H, Xie W, et al. Hydrochar and microplastics disturb soil dissolved organic matter and prominently mitigate ammonia volatilization from wheat growing soil. Applied Soil Ecology. 2022;**178**:104552. DOI: 10.1016/j.apsoil.2022.104552

[45] Lian J, Wu J, Xiong H, Zeb A, Yang T, Su X, et al. Impact of polystyrene nanoplastics (PSNPs) on seed germination and seedling growth of wheat (Triticum aestivum L.). Journal of Hazardous Materials. 2020;**385**:121620. DOI: 10.1016/j.jhazmat.2019.121620

[46] Meng F, Harkes P, van Steenbrugge JJM, Geissen V. Effects of microplastics on common bean rhizosphere bacterial communities. Applied Soil Ecology. 2023;**181**:104649. DOI: 10.1016/j.apsoil.2022.104649

[47] Gao H, Liu Q, Yan C, Mancl K, Gong D, He J, et al. Macro-and/or microplastics as an emerging threat effect crop growth and soil health. Resources, Conservation and Recycling. 2022;**186**:106549. DOI: 10.1016/j.resconrec.2022.106549

[48] Liu J, Wang P, Wang Y, Zhang Y, Xu T, Zhang Y, et al. Negative effects of poly(butylene adipate-co-terephthalate) microplastics on Arabidopsis and its root-associated microbiome. Journal of Hazardous Materials. 2022;**437**:129294. DOI: 10.1016/j.jhazmat.2022.129294

[49] De Souza Machado AA, Lau CW, Kloas W, Bergmann J, Bachelier JB, Faltin E, et al. Microplastics can change soil properties and affect plant performance. Environmental Science & Technology. 2019;**53**(10):6044-6052. DOI: 10.1021/acs.est.9b01339

[50] Lozano YM, Lehnert T, Linck LT, Lehmann A, Rillig MC. Microplastic shape, polymer type, and concentration affect soil properties and plant biomass. Frontiers in Plant Science. 2021;**12**:169. DOI: 10.3389/fpls.2021.616645

[51] Wang F, Feng X, Liu Y, Adams CA, Sun Y, Zhang S. Micro(nano)plastics and terrestrial plants: Up-to-date knowledge on uptake, translocation, and phytotoxicity. Resources, Conservation and Recycling. 2022;**185**:106503. DOI: 10.1016/J.resconrec.2022.106503

[52] Li J, Yu S, Yu Y, Xu M. Effects of microplastics on higher plants: A review. Bulletin of Environmental Contamination and Toxicology. 2022;**109**(2):241-265. DOI: 10.1007/s00128-022-03566-8

[53] Sajjad M, Huang Q, Khan S, Khan MA, Liu Y, Wang J, et al. Microplastics in the soil environment: A critical review. Environmental Technology and Innovation. 2022;**27**:102408. DOI: 10.1016/j.eti.2022.102408

[54] Liu Y, Huang Q, Hu W, Qin J, Zheng Y, Wang J, et al. Effects of plastic mulch film residues on soil-microbe-plant systems under different

soil pH conditions. Chemosphere. 2021;**267**:128901. DOI: 10.1016/j.chemosphere.2020.128901

[55] Menicagli V, Balestri E, Biale G, Corti A, La Nasa J, Modugno F, et al. Leached degradation products from beached microplastics: A potential threat to coastal dune plants. Chemosphere. 2022;**303**:135287. DOI: 10.1016/j.chemosphere.2022.135287

[56] Karalija E, Carbó M, Coppi A, Colzi I, Dainelli M, Gašparović M, et al. Interplay of plastic pollution with algae and plants: Hidden danger or a blessing? Journal of Hazardous Materials. 2022;**438**:129450. DOI: 10.1016/j.jhazmat.2022.129450

[57] Teng L, Zhu Y, Li H, Song X, Shi L. The phytotoxicity of microplastics to the photosynthetic performance and transcriptome profiling of Nicotiana tabacum seedlings. Ecotoxicology and Environmental Safety. 2022;**231**:113155. DOI: 10.1016/j.ecoenv.2021.113155

[58] Bosker T, Bouwman LJ, Brun NR, Behrens P, Vijver MG. Microplastics accumulate on pores in seed capsule and delay germination and root growth of the terrestrial vascular plant Lepidium sativum. Chemosphere. 2019;**226**:774-781. DOI: 10.1016/j.chemosphere.2019.03.163

[59] Sun XD, Yuan XZ, Jia Y, Feng LJ, Zhu FP, Dong SS, et al. Differentially charged nanoplastics demonstrate distinct accumulation in Arabidopsis thaliana. Nature Nanotechnology. 2020;**15**(9):755-760. DOI: 10.1038/s41565-020-0707-4

[60] Sun H, Lei C, Xu J, Li R. Foliar uptake and leaf-to-root translocation of nanoplastics with different coating charge in maize plants. Journal of Hazardous Materials. 2021;**416**:125854. DOI: 10.1016/j.jhazmat.2021.125854

[61] Pehlivan N, Gedik K. Coping with the un-natural: Tracking transcriptional activation and macromolecular profiles in Arabidopsis under microplastic exposure. Environmental and Experimental Botany. 2022;**199**:104870. DOI: 10.1016/j.envexpbot.2022.104870

[62] López MD, Toro MT, Riveros G, Illanes M, Noriega F, Schoebitz M, et al. Brassica sprouts exposed to microplastics: Effects on phytochemical constituents. Science of the Total Environment. 2022;**823**:153796. DOI: 10.1016/j.scitotenv.2022.153796

[63] Gentili R, Quaglini L, Cardarelli E, Caronni S, Montagnani C, Citterio S. Toxic impact of soil microplastics (PVC) on two weeds: Changes in growth, phenology and photosynthesis efficiency. Agron. 2022;**12**(5):1219. DOI: 10.3390/agronomy12051219

[64] Bouaicha O, Tiziani R, Maver M, Lucini L, Miras-Moreno B, Zhang L, et al. Plant species-specific impact of polyethylene microspheres on seedling growth and the metabolome. Science of the Total Environment. 2022;**20**(840):-156678. DOI: 10.1016/j.scitotenv.2022.156678

[65] Menicagli V, Ruffini Castiglione M, Balestri E, Giorgetti L, Sorce C, Spanò C, et al. Early evidence of the impacts of microplastic and nanoplastic pollution on the growth and physiology of the seagrass Cymodocea nodosa. Science of the Total Environment. 2022;**10**(3):438156514. DOI: 10.1016/j.scitotenv.2022.156514

[66] Pinto-Poblete A, Retamal-Salgado J, López MD, Zapata N, Sierra-Almeida A, Schoebitz M. Combined effect of microplastics and Cd alters the enzymatic activity of soil and the productivity of strawberry plants.

Plants. 2022;**11**(4):536. DOI: 10.3390/plants11040536

[67] Lian Y, Liu W, Shi R, Zeb A, Wang Q, Li J, et al. Effects of polyethylene and polylactic acid microplastics on plant growth and bacterial community in the soil. Journal of Hazardous Material. 2022;**435**:129057. DOI: 10.1016/j.jhazmat.2022.129057

[68] Yu G, Huang S, Luo X, Zhao W, Zheng Z. Single and combined toxicity effects of nanoplastics and bisphenol F on submerged the macrophyte Hydrilla verticillata. Science of the Total Environment. 2022;**814**:152564. DOI: 10.1016/j.scitotenv.2021.152564

[69] Xu Z, Zhang Y, Lin L, Wang L, Sun W, et al. Toxic effects of microplastics in plants depend more by their surface functional groups than just accumulation contents. Science of the Total Environment. 2022;**833**(10):155097. DOI: 10.1016/j.scitotenv.2022.155097

[70] Zhang Y, Zhang C, Jiang M, Zhou G. Bio-effects of bio-based and fossil-based microplastics: Case study with lettuce-soil system. Environmental Pollution. 2022;**306**:119395. DOI: 10.1016/j.envpol.2022.119395

[71] Yang X, Liao H, Tan A, Gan S, Yang GL. Effects of microplastics on heavy metal stress in duckweed. SSRN. 2022:4171706-4171726. DOI: 10.2139/ssrn.4171706

[72] Rozman U, Kokalj A, Dolar A, Drobne D, Kalčíková G. Long-term interactions between microplastics and floating macrophyte Lemna minor: The potential for phytoremediation of microplastics in the aquatic. Science of the Total Environment. 2022;**831**:154866-154875. DOI: 10.1016/j.scitotenv.2022.154866

[73] De Silva YSK, Rajagopalan UM, Kadono H, Li D. Effects of microplastics on lentil (Lens culinaris) seed germination and seedling growth. Chemosphere. 2022;**2**:135162. DOI: 10.1016/j.chemosphere.2022.135162

[74] Shi R, Liu W, Lian Y, Wang Q, Zeb A, Tang J. Phytotoxicity of polystyrene, polyethylene and polypropylene microplastics on tomato (Lycopersicon esculentum L.). Journal of Environmental Management. 2022;**317**:115441-115452. DOI: 10.1016/j.jenvman.2022.115441

[75] Yang C, Gao X. Impact of microplastics from polyethylene and biodegradable mulch films on rice (*Oryza sativa* L.). Science of the Total Environment. 2022;**828**:154579-154589. DOI: 10.1016/j.scitotenv.2022.154579

[76] Wang J, Lu S, Bian H, Xu M, Zhu W, Wang H, et al. Effects of individual and combined polystyrene nanoplastics and phenanthrene on the enzymology, physiology, and transcriptome parameters of rice (*Oryza sativa* L.). Chemosphere. 2022;**304**:135341. DOI: 10.1016/j.chemosphere.2022.135341

[77] Tan J, Chen Y, Mo Z, Tan C, Wen R, Chen Z, et al. Zinc oxide nanoparticles and polyethylene microplastics affect the growth, physiological and biochemical attributes, and Zn accumulation of rice seedlings. Environmental Science and Pollution Research. 2022;**29**:61534-61546. DOI: 10.1007/s11356-022-19262-3

[78] Ma J, Aqeel M, Khalid N, Nazir A, Alzuaibr FM, Al-Mushhin A, et al. Effects of microplastics on growth and metabolism of rice (*Oryza sativa* L.). Chemosphere, 307, 2022:(1):135749-61. DOI: 10.1016/j.chemosphere.2022.135749

[79] Zhang L, Cheng Y, Qian Y, Ding T, Li J. Phytotoxicity and accumulation of BPS to Pistia stratiotes under the influence of microplastics. Chemosphere.

2022;**307**:135854. DOI: 10.1016/j.chemosphere.2022.135854

[80] Cui M, Yu S, Yu Y, Chen X, Li J. Responses of cherry radish to different types of microplastics in the presence of oxytetracycline. Plant Physio Biochem. 2022;**191**:1-9. DOI: 10.1016/j.plaphy.2022.09.012

[81] Weerasinghe WMTM, Madawala HMSP. Potential impacts of two types of microplastics on (Solanum lycopersicum L.) and arbuscular mycorrhizal fungi. Ceylon Journal of Science. 2022;**51**(2):137-146. DOI: 10.4038/cjs.v51i2.8008

[82] Zou J, Wang C, Li J, Wei J, Liu Y, Hu L, et al. Effect of polyethylene (LDPE) microplastic on remediation of cadmium contaminated soil by Solanum nigrum L. Journal of Geosciences and Environmental Protocol. 2022;**10**(1):49-64. DOI: 10.4236/gep.2022.101004

[83] Singh B, Singh K. Microplastics contamination in soil affects growth and root nodulation of fenugreek (Trigonella foenum-graecum L.) and 16 s rRNA sequencing of rhizosphere. Journal of Hazardous Materials. 2022;7:100146-100154. DOI: 10.1016/j.hazadv.2022.100146

[84] Ren X, Wang L, Tang J, Sun H, Giesy JP. Combined effects of degradable film fragments and micro/nanoplastics on growth of wheat seedling and rhizosphere microbes. Environmental Pollution. 2022;**294**:118516-118526. DOI: 10.1016/j.envpol.2021.118516

[85] Lee T, Kim L, Kim D, An S, An YJ. Microplastics from shoe sole fragments cause oxidative stress in a plant (Vigna radiata) and impair soil environment. Journal of Hazardous Materials. 2022;**429**:128306-128314. DOI: 10.1016/j.jhazmat.2022.128306

[86] Brown R, Chadwick D, Zang H, Graf M, Liu X, Wang K, et al. Bioplastic (PHBV) addition to soil alters microbial community structure and negatively affects plant-microbial metabolic functioning in maize. Journal of Hazardous Materials. 2023;**441**:129959-129971. DOI: 10.1016/j.jhazmat.2022.129959

[87] Ingraffia R, Amato G, Iovino M, Rillig MC, Giambalvo D, Frenda AS. Polyester microplastic fibers in soil increase nitrogen loss via leaching and decrease plant biomass production and N uptake. Environmental Research Letters. 2022;**17**(5):054012. DOI: 10.1088/1748-9326/ac652d

[88] Kaphle A, Navya PN, Umapathi A, Daima HK. Nanomaterials for agriculture, food and environment: Applications, toxicity and regulation. Environmental Chemistry Letters. 2017;**16**(1):43-58. DOI: 10.1007/s10311-017-0662-y

[89] Wang F, Zhang X, Zhang S, Zhang S, Sun Y. Interactions of microplastics and cadmium on plant growth and arbuscular mycorrhizal fungal communities in an agricultural soil. Chemosphere. 2020;**254**:126791. DOI: 10.1016/j.chemosphere.2020.126791

[90] Colzi I, Renna L, Bianchi E, Castellani MB, Coppi A, Pignattelli S, et al. Impact of microplastics on growth, photosynthesis and essential elements in Cucurbita pepo L. Journal of Hazardous Materials. 2022;**423**:127238. DOI: 10.1016/j.jhazmat.2021.127238

[91] Zong X, Zhang J, Zhu J, Zhang L, Jiang L, Yin Y, et al. Effects of polystyrene microplastic on uptake and toxicity of copper and cadmium in hydroponic wheat seedlings (Triticum aestivum L.). Ecotoxicology and Environmental Safety. 2021;**217**:112217. DOI: 10.1016/j.ecoenv.2021.112217

[92] Li L, Luo Y, Li R, Zhou Q, Peijnenburg WJGM, Yin N, et al. Effective uptake of submicrometre plastics by crop plants via a crack-entry mode. Nature Sustainability. 2020;**3**(11):929-937. DOI: 10.1038/s41893-020-0567-9

[93] Jiang X, Chen H, Liao Y, Ye Z, Li M, Klobučar G. Ecotoxicity and genotoxicity of polystyrene microplastics on higher plant Vicia faba. Environmental Pollution. 2019;**250**:831-838. DOI: 10.1016/j.envpol.2019.04.055

[94] Dong Y, Gao M, Qiu W, Song Z. Uptake of microplastics by carrots in presence of As (III): Combined toxic effects. Journal of Hazardous Materials. 2021;**411**:125055. DOI: 10.1016/j.jhazmat.2021.125055

[95] Mamathaxim N, Song W, Wang Y, Habibul N. Effects of microplastics on arsenic uptake and distribution in Rice seedlings. SSRN Electronic Journal. 2022;**862**:4202084. DOI: 10.2139/SSRN.4202084

[96] Capaldi Arruda SC, Diniz Silva AL, Moretto Galazzi R, Antunes Azevedo R, Zezzi Arruda MA. Nanoparticles applied to plant science: A review. Talanta. 2015;**131**:693-705. DOI: 10.1016/j.talanta.2014.08.050

[97] Verma V, Ryan KM, Padrela L. Production and isolation of pharmaceutical drug nanoparticles. International Journal of Pharmaceutics. 2021;**603**(March):120708. DOI: 10.1016/j.ijpharm.2021.120708

[98] Khan N, Bano A. Role of plant growth promoting rhizobacteria and Ag-nano particle in the bioremediation of heavy metals and maize growth under municipal wastewater irrigation. International Journal of Phytoremedicine. 2016;**18**(3):211-221. DOI: 10.1080/15226514.2015.1064352

[99] Goswami P, Yadav S, Mathur J. Positive and negative effects of nanoparticles on plants and their applications in agriculture. Plant Science Today. 2019;**6**(2):232-242. DOI: 10.14719/pst.2019.6.2.502

[100] El-Moneim DA, Dawood MFA, Moursi YS, Farghaly AA, Afifi M, Sallam A. Positive and negative effects of nanoparticles on agricultural crops. Nanotechnology and Environmental Engineering. 2021;**6**(2):1-11. DOI: 10.1007/s41204-021-00117-0

[101] Fatemi H, Esmaiel Pour B, Rizwan M. Foliar application of silicon nanoparticles affected the growth, vitamin C, flavonoid, and antioxidant enzyme activities of coriander (Coriandrum sativum L.) plants grown in lead (Pb)-spiked soil. Environmental Science and Pollution Research. 2021;**28**(2):1417-1425. DOI: 10.1007/s11356-020-10549-x

[102] Khan M, Khan AU, Hasan MA, Yadav KK, Pinto MMC, Malik N, et al. Agro-nanotechnology as an emerging field: A novel sustainable approach for improving plant growth by reducing biotic stress. Applied Sciences. 2021;**11**(5):2282. DOI: 10.3390/app11052282

[103] Birbaum K, Brogioli R, Schellenberg M, Martinoia E, Stark WJ, Günther D, et al. No evidence for cerium dioxide nanoparticle translocation in maize plants. Environmental Science & Technology. 2010;**44**(22):8718-8723. DOI: 10.1021/es101685f

[104] Rehman AU, Nazir S, Irshad R, Tahir K, ur Rehman K, Islam RU, et al. Toxicity of heavy metals in plants and animals and their uptake by magnetic iron oxide nanoparticles. Journal of Molecular Liquids 2021;321, 114455. DOI: 10.1016/j.molliq.2020.114455

[105] Bhagat Y, Gangadhara K, Rabinal C, Chaudhari G, Ugale P. Nanotechnology in agriculture: A review. Journal of Pure and Applied Microbiology;**9**:737-747

[106] Ma X, Yan J. Plant uptake and accumulation of engineered metallic nanoparticles from lab to field conditions. Current Opinion in Environmental Sciences and Health. 2018;**6**:16-20. DOI: 10.1016/j.coesh.2018.07.008

[107] Ruttkay-Nedecky B, Krystofova O, Nejdl L, Adam V. Nanoparticles based on essential metals and their phytotoxicity. Journal of Nanobiotechnology. 2017;**15**(1):1-19. DOI: 10.1186/s12951-017-0268-3

[108] Lin D, Xing B. Root uptake and phytotoxicity of ZnO nanoparticles. Environmental Science & Technology. 2008;**42**(15):5580-5585. DOI: 10.1021/es800422x

[109] Pérez-Labrada F, Hernández-Hernández H, López-Pérez MC, González-Morales S, Benavides-Mendoza A, Juárez-Maldonado A. Nanoparticles in plants: Morphophysiological, biochemical, and molecular responses. In: Plant Life under Changing Environment Responses and Management. Academic Press; 2020. pp. 289-322. DOI: 10.1016/b978-0-12-818204-8.00016-3

[110] Ayyaz A, Fang R, Ma J, Hannan F, Huang Q, Athar H ur R, et al. Calcium nanoparticles (Ca-NPs) improve drought stress tolerance in Brassica napus by modulating the photosystem II, nutrient acquisition and antioxidant performance. NanoImpact. 2022;28:100423. DOI: 10.1016/j.impact.2022.100423

[111] Roy T, Dey TK, Jamal M. Microplastic/nanoplastic toxicity in plants: An imminent concern. Environmental Monitoring and Assessment. 2022;**195**(1):1-35. DOI: 10.1007/s10661-022-10654-z

[112] Tripathi D, Singh M, Pandey-Rai S. Crosstalk of nanoparticles and phytohormones regulate plant growth and metabolism under abiotic and biotic stress. Plant Stress. 2022;**6**:100107-100118. DOI: 10.1016/j.stress.2022.100107

Chapter 2

Perspective Chapter: Regulatory Network in Plant under Abiotic Stress

Noreen Fatima, Faiza Khan and Asif Saeed

Abstract

Abiotic stress is one of the primary causes of crop yield loss worldwide; it contributes to a nearly 50% reduction in crop production. Anti-oxidants are produced in plants to scavenge ROS that causes cellular damage during abiotic stress. The plant stress response is a complex mechanism that involves protein initiation pathways, abscisic acid signaling, transcriptional and post-transcriptional modifications. Abiotic stress-inducible genes are divided into two groups based on protein products, one is for signal transduction and other is for expressing resistance. Transcriptional factors bind to the promoter of the target gene at specific DNA sequences thus regulating the gene expression, so different kinds of transcriptional factors known as regulons are involved in regulation of genes during drought, salinity and cold stress.

Keywords: abiotic stress, transcriptional factors, signalling, genes

1. Introduction

Abiotic stress is one of the main reasons for crop yield loss worldwide, almost more than 50% reduction in crop yield is due to abiotic stress. Plant are sessile in nature and environmental conditions constantly affect the plant. When these conditions are extreme and rapid, plants face stress conditions. But stresses are not necessarily a problem because plants have defense mechanisms to reduce or avoid the damage caused by these stresses.

For salinity and drought, it is very easy to distinguish the primary stress signals from the secondary stress signals, which is caused by too much salt or too little water. Primary signals caused by the drought which is hyperosmotic stress. Drought is often callas as osmotic stress because hypo-osmotic condition does not harmful for the cell Salt stress cause both the osmotic can ionic effects on cells. The secondary effects of salinity and drought cause the oxidative damage to cell like damage to cellular proteins, lipids, nucleic acid and dysfunction of metabolites. Salinity and drought have overlapping and unique stress signals. Both cause the hyperosmotic signals which causes the phyto-hormone abscisic acid accumulation [1].

Abiotic stress such as high temperature, low temperature, high light intensity, heavy metals, osmotic stress and the number of herbicides and toxic elements leads

Stress	Plant response	Effects of stress on plant
Drought stress	Stomata closure, leaf rolling, accumulation of metabolites	Reduction in photosynthesis due to inability of water transport to leaf
Heat stress	Protein stability support survival, activation of protein repair systems	High evaporation, water deficit, plant death due to enzyme turnover
Cold stress	Synthesis of hydrophilic protein and accumulation of osmolytes to prevent ice crystal formation	Slow rate of biochemical reactions, photosynthesis, leading to oxygen radical damage and CO_2 fixation lags. Formation of ice crystal which disrupt the cell membrane

Table 1.
Effects of abiotic stress and response of plant.

to a high amount of reactive oxygen species (ROS) productions that cause cellular damage and reduce the amount of photosynthesis in the plant (**Table 1**). The plant produces the antioxidants that remove the ROS but sometimes in extreme stress conditions amount of ROS is so much high that antioxidants are unable to remove them. So a high amount of ROS in the cell leads to cell damage that reduces the various metabolic processes of the plant. To repair these damages, plant cells have a complex defense system, including the antioxidant stress-related defense genes that cause the changes in plant cell machinery.

Stress-responsive pathways in plants are not linear, they are complex that involved multiple pathways and specific tissues and cellular compartments [2]. When plants face stress, at the initial stress stages, the activation of protein signaling pathways and activation of Ca^+ act as the initial sensor that leads to the expression of stress-responsive genes and physiological changes [3]. Accumulation of abscisic acid (ABA) in cell plays a crucial role in transduction pathways, and stress signaling and activate many defense responses [4].

It is known that abiotic stress, through regulation of protein turnover and gene expression, changes the abundance of many proteins and mRNAs, showing that transcriptional and post-transcriptional gene regulation plays a very important role in cellular function adaptation to environmental changes. Recent advances in genomics, metabolomics and proteomics have provided a way to study the gene regulatory network system which consists of inducible genes, expression programming and regulatory elements [5]. Through genetic studies, scientists revealed that stress-related attributes are quantitative trait loci, so the selection of these traits is difficult.

The plant produces various metabolic proteins such as osmoprotectants, regulatory proteins that play role in signal transduction pathways like kinases and transcriptional factors in responses to abiotic stress. Transcriptional gene regulation is controlled by transcriptional factor binding sites (TFBS) and a network of transcriptional factors [6].

Transcriptional factors are proteins in nature that have DNA binding domains that bind to cis-acting elements that are present in the target gen promoter. TFs repress (repressor) or induce (activators) the RNA polymerase activity that leads to the expression of gene regulation. TFs are grouped into families based on their DNA binding domain [7]. The absence and presence of TFs (activators or repressors) lead to the regulation of gene transcription that involved the whole signaling cascade events specified by plant developmental stage, environmental condition and tissue type [8].

Abiotic stress-inducible genes are divided into two groups based on protein products. The first group of genes, proteins of these genes play role in the regulation of signal transduction such as TFs and gene expression, another group of genes whose proteins directly confer resistance in plant stress against environmental stress such as anti-freezing protein, late embryogenesis abundant (LEA), enzymes for synthesis of proline, betaine and osmoregulators. Four different regulons have been identified, two regulons are ABA dependent (1 and 2) and two are ABA independent (3 and 4): (1) AREB (ABA-responsive element-binding protein) /ABF regulons (ABA-binding factor); (2) MYC/MYB (myelocytomatosis oncogene/myeloblastosis oncogene) regulons; (3) CBF/DREB regulons (4) ZF-HD (zinc-finger homeodomain) and NAC (ATAF, CUC and NAM) regulons.

Our knowledge related to molecular mechanisms involved in plant defense against abiotic stress is limited but many numbers of genes have been identified in recent years that are involved in these responses. These genes can be induced by stimuli and proteins that confer resistance to abiotic conditions.

The book chapter aims to report recent advances in abiotic stress-responsive mechanisms and describe the regulation of gene expression.

2. Abiotic stress-inducible genes

Many genes are involved in plant response to abiotic stress. The study of the function of abiotic stress-inducible genes is an important tool to study the molecular mechanisms of stress tolerance and plant responses but it also helps scientists to improve the stress tolerance in crop plats by manipulation of genes. Hundreds of genes have been identified that play role in abiotic stress responses [9]. Most of the drought-inducible genes are induced by cold stress and salinity stress.

These genes are divided into 3 groups: (1) in this group, genes product directly protect the cell against stresses e.g. heat shock proteins (HSPs), chaperones, osmoprotectants, LEA proteins, detoxification of enzymes, antifreeze proteins and free radical scavengers [10]; (2) in this group, genes that involved in signaling cascade and transcriptional control e.g. Calcium-dependent protein kinase (CDPK) and Mitogen-activated protein kinase (MAPK), phospholipases and transcriptional factors [11]; (3) genes that are involved in ion and water uptake and transport via ion transporter and aquaporin [12].

3. Transcriptional factors genes

Plant growth and developmental processes are affected by environmental stress. These stress are salinity, drought and high temperature etc. Human activities cause adverse effects on the atmosphere that leads to an increase in conc. of O_3 in the troposphere and this causes oxidative stress, oxidative stress cause the destruction of various important proteins and cells, and a reduction in crop yield.

Susceptibility or tolerance is a very complex phenomenon because stress can occur at any plant developmental stage and mostly more than one stress affects the plant [13]. Abiotic stress responses involved the production of metabolic proteins such as proteins which are involved in the production of regulatory proteins and osmoprotectants that play role in signal transduction e.g. transcription factors or kinases.

Regulation of responses needs proteins that operate in signal transduction pathways, e.g. transcription factors. These factors bind to the promoter of the target gene at specific DNA sequences thus regulating the gene expression. This type of regulation is known as regulons. Four different regulons have been identified that are active in plants during abiotic stress. Dehydration responsive element binding protein/C-repeat binding factor (DREB1/CBF), DREB2 regulons play important role in ABA-independent gene expression, but in ABA-dependent gene expression regulons ABA-responsive element binding protein/ABREF binding factor (AREB/ABF) [14]. Some other regulons like NAC (No apical meristem) and MYB/MYC (Myeloblastosis-Myelosytomatosis) are also involved in gene expression of the abiotic stress-responsive gene. TF OsNAC6 is a NAC-type in produced by the plant during abiotic stress especially cold, salinity and drought. In Rice, over-expression of OsNAC6 during abiotic stress is studied by using microarray analysis [15]. Transcription factors can be the target for genetic engineering for the development of abiotic stress-resistant crop plants.

4. Transcription factor for drought stress

The plant genome controls the regulation of TF under water deficit conditions. By using the microarray technique in *Arabidopsis thaliana*, also known as the model plant, many genes have been identified that are repressed or induced in response to a condition that leads to cellular drought stress [16]. The drought genes have been divided into four groups, transcriptional regulations, protection of cellular structures, cellular metabolism and signal transduction. Almost six different classes of TFs have been identified that participate in gene repression or induction under drought conditions. Multiple treatments that mimic drought stress, induce NAC domain and Homebox domain containing TFs. Adaptation of plants to water stress is promoted by the accumulation of proteins which have structural and metabolic functions. Late embryogenesis abundant (*Lea*) gene play role in plant protection. In desiccating seeds, *Lea* genes are expressed. *Lea* genes produce a hydrophilic protein that protects the plant membranes and proteins because of its chaperons-like function. In rice plants, these hydrophilic proteins are identified that protect the cell membrane from injury under drought stress [17].

5. Transcriptional factors in drought stress

In wheat genome microarray analysis was performed, and the result of the analysis showed that in 300 unique ESTs (expressed sequence tag), 18% of genes were down-regulated and 30% of genes were up-regulated under water stress [18]. Bray [19] reported the 130 genes in *A. thaliana* that were up-regulated under water stress. Under water stress, these genes are involved in cellular response by detoxification, signaling events and other functions.

5.1 Salinity stress

Large world cultivated areas are salt-affected. Salt stress affects plant growth, development, metabolic processes and physiological processes that lead to a reduction in crop yield [20]. Plant water relations are also affected by salt stress. Salinity creates ionic stress by the accumulation of Cl^- and Na^+ ions in the cell. Homeostasis of some other initial ions e.g. Ca^{++}, NO^{-3} and K^+. The plant cell membranes are also affected by salt stress. It affects the lipid and proteins of the cell membrane which leads to cell injury because of hyperosmotic stress and ion imbalance. These changes in plants cause disturbance in normal plant development and growth [21, 22]. High NaCl concentration in the root zone affects the plant metabolism and physiology at various levels (oxidative stress, water deficit, nutrient imbalance and ion toxicity [23]. At the initial stages of salinity stress, the plant experienced drought stress which causes a reduction in leaf expansion. When exposed to long-term salinity stress, ionic stress occurs in the plant that causes the premature senescence of leaves so that the photosynthetic area is reduced.

The salt-responsive gene are classified into two groups: 1) early responsive genes and 2) delay responsive genes [24]. The early responsive genes are induced transiently and quickly while the late responsive genes are activated slowly and the expression of these genes is constant. Early responsive genes produce proteins that are transcription factors. These transcription factors are involved in the activation of downstream late response genes [1]. When microarray expression profile, a T-DNA insertion knockout mutant of AtNHX1 and rescued line (NHX1::nhx1) exposed to both long (1 and 2 weeks) and short (12 h and 48 h) duration of salt stress were studied in wild type plant, 147 transcripts were found which showed both significant effects of AtNHX1 and salt responsiveness.

A large number of genes have been identified that are involved in responses conferring salt tolerance. As functional components, many genes are identified in plant response under stress conditions, including genes that code for an enzyme that has a role in detoxification like glutathione peroxidase [25], osmolytes such as LEA and glycine-betaine [26], AtNHX1 Na^+/K^+ antiporter [27], Alfin1 TF [28] and AtHAL3 flavoprotein [29]. The complete study of transcriptomes showed that synthesis of ion transporter and osmolytes and regulation of translational and transcriptional machinery have important roles in salinity stress response. Induction of various transcripts that encode for RNA-binding protein, specific TFs, elongation factors, ribosomal genes and translation initiation is reported and these transcripts are important during salinity stress [30].

Some stress-specific consensus sequences are identified in the promoter of genes, these regions help in activating or repressing these gene transcriptions, TF should be located in the nucleus that binds with DNA and interacts with the basal transcription apparatus. TFs that play role in stress responses are DRE-related binding factors, putative Zn-finger proteins, leucine zipper DNA binding protein, AP2/EREBP, myb proteins and bZIP/HD-ZIPs [31], Interact with osmotic regulated genes promoter [32]. AP2/ERF protein domain has CBF or DREB proteins that bind with DRE (dehydration response elements) or C-repeats. DRE/C-repeat promoter sequence and DREB/CBF factors in stress-activated genes control the expression of salt-responsive genes. In Arabidopsis, various stress-inducible genes e.g. Kin1, Cor6.6, rd29A and Cor15a are the target of DREB/CBFs and contain DRE/C-repeat sequences in the promoter region.

Basic leucine zipper protein consists of a DNA binding domain that binds to the ACGT core sequence of the target region. One subfamily of bZIP has been identified that is linked with ABA response ABRE binding factors (ABFs/AREBs), ABI5 and its homologs. Under salt stress when plants go under dehydration ABRE binding factors are transcribed by the plant cell [33].

Some regulatory intermediates such as SOS3 (calcium-binding protein), SOS2 (Suc nonfermenting -like) kinase, mitogen-activated protein kinase and Calcium-dependent protein kinases [34]. Salt tolerance and ion homeostasis are regulated by a single pathway which includes SOS1, SOS3 and SOS2, their functions are calcium-dependent. SOS1 encode the protein (plasma membrane Na^+/H^+ antiporter) which play important role in Na^+ extrusion [35]. This antiporter makes a component which increases the Ca^{++} and reversible phosphorylation [36].

Several studies reported that oxidative stress and reactive oxygen species (ROS) may be mediating the toxic effects of NaCl in legumes and vascular plants [37]. ROS are produced in plant chloroplast by direct excitation energy transfer from chlorophyll to produce the singlet oxygen or in Mehler reaction by the reduction of univalent oxygen at photosystem 1 and in mitochondria. ROS can interact with many cellular components and trigger the peroxidation reaction which leads to significant damage to cellular lipids, proteins and nucleic acids. Plant have an antioxidant defense system that helps in the detoxification of ROS and maintain redox homeostasis. This system consists of enzymes and non-enzyme components which maintain the level of ROS in the plant cells. Major non-enzymatic antioxidants present in plants are ascorbic acid and glutathione, some others are tocopherol, alkaloids flavonoids and carotenoids.

5.2 Cold or chilling stress

Cold stress inhibits the expression of the full genetic potential of the crop plants which leads to a reduction in metabolic reaction and indirectly through the oxidative, cold-induced osmotic (chilling induced prevent the water uptake and freezing induced leads to cell dehydration), and other stresses. Cold stress has 2 types: 1) chilling stress at <20°C and freezing stress at <0°C. These low temperatures cause adverse effect on plant cells and growth.

Chilling and cold stress leads to slow the biochemical process such as effect on enzyme and membrane transport. It also leads to the formation of ice crystals which causes the cell membrane system disruption [38]. A large number of genes studies have been used for the identification of genes that respond to cold stress in Arabidopsis. Results of the studies showed that under cold stress, mRNA levels of many genes that are important for plant growth and development altered [39].

The cold induction of genes involved in lipid signaling, calcium signaling or encoding receptor-like protein kinases are affected by the ice1 mutation. TFs that bind with DRE/CRT is known as DREB2 and DREB1/CTR-binding factors. AP_2/ERF family of transcription factors increases under cold stress. This family have CBFs which bind to the promoter region of COR genes and activate the expression of the genes. CBFs regulate the expression of many genes which are involved in ROS detoxification, transcription, phosphoinositide metabolism, membrane transport, signaling, osmolytes and hormone metabolism [40].

The first cDNA isolated from Arabidopsis, encodes for the DRE binding proteins (DREB1A and DREB2A) reported in scientific study [41] after that many *DREB* genes were isolated from many plants. In barley and wheat, the number of *CBF* homologs have been identified and mapped to low temperate QTLs, and *Fr-2* chromosomal region [42]. Results of the scientific research showed that DREB1/CBF regulons are ubiquitous in nature within higher plants. Expression of *DREB1* genes was investigated in many crop plants under different abiotic stresses. It was concluded that *AtDREB1* gene expression is induced by only cold stress, not due to dehydration or salinity [43]. Another gene *CBF* gene expression was studied and concluded that this gene also expressed under sold stress and its mRNA was detected after 30 minutes in the plant when exposed to 4°C and showed the maximum expression at 1 hour [44]. Two cold-responsive TFs of genes RAP2.1 and RAP2.7 can sub-regulate the CBF region. This results showed by the microarray analysis when performed on transgenic Arabidopsis ectopically expressing the CBFs. CBFs can regulate the expression of some COR genes that create the cold resistance in plants.

The expression of ZAT10 can be induced by the CBFs which can down regulate the expression of COR genes. Phosphorylation and sumoylation which is induced by cold stress can activate the constitutive expressed ICE1. Activated ICE1 can induced the transcription of reprime *MYP15* and *CBFs*. MYB15 and ZAT12 can negatively regulate the expression of *CBFs*. HOS1 mediate the proteolysis and ubiquitination of ICE1, so that it negatively regulate the CBF regulons [45]. Small ubiquitin-related Modifier proteins (SUMO) can induced the sumoylation. Sumoylation protect the ubiquitination of the proyein because it protect the protein from the proteosomal degeradation [46].

Key findings

Late embryogenesis abundant (Lea) gene play role in plant protection during drought and cold stress, Lea genes produce a hydrophilic protein. In rice plants, these hydrophilic proteins are identified that protect the cell membrane from injury under drought stress. Under salt stress when plants go under dehydration, ABRE binding factors are transcribed by the plant cell which protect the plant against salinity. A large number of genes studies have been used for the identification of genes that respond to cold stress in Arabidopsis. Results of the studies showed that under cold stress, mRNA levels of many genes that are important for plant growth and development altered.

Author details

Noreen Fatima*, Faiza Khan and Asif Saeed
University of Agriculture, Faisalabad, Punjab, Pakistan

*Address all correspondence to: norfatima44@yahoo.com

References

[1] Zhu JK. Salt and drought stress signal transduction in plants. Annual Review of Plant Biology. 2002;**53**:247-273

[2] Dombrowski JE. Salt stress activation of wound-related genes in tomato plants. Plant Physiology. 2003;**132**:2098-2107

[3] Bressan RA, Hasegawa PM, Pardo M. Plants use calcium to resolve salt stress. Trends in Plant Science. 1998;**3**:411-412

[4] Wasilewska A, Vlad F, Sirichandra C, et al. An update on abscisic acid signaling in plants and more. Molecular Plant. 2008;**1**:198-217

[5] Munns R. Genes and salt tolerance: Bringing them together. New Phytologyst. 2005;**167**:645-663

[6] Chaves MM, Oliveira MM. Mechanisms underlying plant resilience to water deficits: Prospects for water-saving agriculture. Journal of Experimental Botany. 2004;**55**(407):2365-2384

[7] Riechmann JL, Heard J, Martin G. Arabidopsis transcription factors: Genomewide comparative analysis among eukaryotes. Science. 2000;**290**(5499):2105-2110

[8] Wyrick JJ, Young RA. Deciphering gene expression regulatory network. Current Opinion in Genetic and Development. 2002;**12**:130-136

[9] Avni ÖH, Eyidoğan F, Selçuk F, Tufan ÖM, da Silva JAT, Yücel M. Revealing response of plants to biotic and abiotic stresses with microarray technology. Genes, Genomes and Genomics. 2008:14-48

[10] Wang WX, Vinocur B, Altman A. Plant responses to drought, salinity and extreme temperatures: Towards genetic engineering for stress tolerance. Planta. 2003;**218**:1-14

[11] Frank W, Munnik T, Kerkmann K, Salamini F, Bartels D. Water deficit triggers phospholipase D activity in the resurrection plant Craterostigma plantagineum. Plant Cell. 2000;**12**:111-124

[12] Blumwald E. Sodium transport and salt tolerance in plants. Current Opinion in Cell Biology. 2000;**12**:431-434

[13] Chinnusamy V, Schumaker H, Zhu J-K. The Arabidopsis LOS5/ABA3 locus encodes a molybdenum cofactor sulfurase and modulates cold stress– And osmotic stress–responsive gene expression. Journal of Experimental Botany. 2004;**55**(395):225-236

[14] Saibo NJM, Lourenco T, Oliveira MM. Transcription factors and regulation of photosynthetic and related metabolism under environmental stresses. Annals of Botany. 2008;**103**:609-623

[15] Nakashima K, Tran LSP, Van Nguyen D, Fujita M, Maruyama K, Todaka D, et al. Functional analysis of NAC-type transcription factor OsNAC6 involved in abiotic and biotic stressresponsive gene expression in rice. The Plant Journal. 2007;**51**:617-630

[16] Seki M, Narusaka M, Ishida J, Nanjo T, Fujita M, Oono Y, et al. Monitoring the expression profiles of 7000 Arabidopsis genes under drought, cold and high-salinity stresses using a full-length cDNA microarray. The Plant Journal. 2002;**31**:279-292

[17] Chandra BR, Jingxian Z, David BL, Hod T-H, Wue R, Nguyenf HT. HVA1,

a LEA gene from barley confers dehydration tolerance in transgenic rice (Oryza sativa L.) via cell membrane protection. Plant Science. 2004;**166**:855-862

[18] Way H, Chapman S, McIntyre L, Casu R, Xue GP, Manners J, et al. Identification of differentially expressed genes in wheat undergoing gradual water deficit stress using a subtractive hybridisation approach. Plant Science. 2005;**168**:661-670

[19] Bray EA. Classification of genes differentially expressed during water-deficit stress in Arabidopsis thaliana: An analysis using microarray and differential expression data. Annals of Botany. 2002;**89**:803-811

[20] Zhang L et al. Identification of an apoplastic protein involved in the initial phase of salt stress response in rice root by twodimensional electrophoresis. Plant Physiology. 2009;**149**:916-928

[21] Fujii H, Zhu J-K. An autophosphorylation site of the protein kinase SOS2 is important for salt tolerance in Arabidopsis. Molecular Plant. 2009;**2**:183-190

[22] López-Pérez L et al. Changes in plasma membrane lipids, aquaporins and proton pump of broccoli roots, as an adaptation mechanism to salinity. Phytochemistry. 2009;**70**:492-500

[23] Vinocur B, Altman A. Recent advances in engineering plant tolerance to abiotic stress: Achievements and limitations. Curr Opin Biotech. 2005;**16**:123-132

[24] Sairam RK, Tyagi A. Physiological and molecular biology of salinity stress tolerance in plants. Current Science. 2004;**86**:407-420

[25] Roxas VP, Smith RK, Allen ER, Jr A, R.D. Overexpression of glutathione Stransferase/glutathione peroxidase enhances the growth of transgenic tobacco seedlings during stress. Nature Biotechnology. 1997;**15**:988-991

[26] Xu D, Duan X, Wang B, Hong B, Ho TDD, Wu R. Expression of a late embryogenesis abundant protein gene, HVA1, from barley confers tolerance to water deficit and salt stress in transgenic rice. Plant Physiology. 1996;**110**:249-257

[27] Apse MP, Aharon GS, Sneddon WA, Blumwald E. Salt tolerance conferred by overexpression of a vacuolar Na+/H+ antiporter in Arabidopsis. Science. 1999;**285**:1256-1258

[28] Bastola DR, Pethe VV, Winicov I. A1fin1, a novel zincfinger protein in alfalfa roots that binds to promoter elements in the salt-inducible MsPRP2 gene. Plant Molecular Biology. 1998;**38**:1123-1135

[29] Espinosa-Ruiz A, Belles JM, Serrana R, Culianez-Macla FA. Arabidopsis thaliana AtHAL3: A flavoprotein related to salt and osmotic tolerance and plant growth. The Plant Journal. 1999;**20**:529-539

[30] Sahi C, Singh A, Blumwald E, Grover A. Beyond osmolytes and transporters: Novel plant salt-stress tolerance-related genes from transcriptional profiling data. Physiologia Plantarum. 2006;**127**:1-9

[31] Chen W, Provart NJ, Glazebrook J, Katagiri F, Chang HS, Eulgem T, et al. Expression profile matrix of Arabidopsis transcription factor genes suggests their putative functions in response to environmental stresses. Plant Cell. 2002;**14**:559-574

[32] Hasegawa PM, Bressan RA, Zhu J-K, Bohnert HJ. Plant cellular and molecular

responses to high salinity. Annual Review of Plant Physiology and Plant Molecular Biology. 2000;**51**:463-499

[33] Uno Y, Furihata T, Abe H, Yoshida R, Shinozaki K, Yamaguchi-Shinozaki K. Arabidopsis basic leucine zipper transcription factors involved in an abscisic aciddependent signal transduction pathway under drought and high-salinity conditions. Proc. Natl. Acad. Sci. USA. 2000;**97**:11632-11637

[34] Halfter U, Ishitani M, Zhu J-K. The Arabidopsis SOS2 protein kinase physically interacts with and is activated by the calcium-binding protein SOS3. Proc Natl Acad Sci USA. 2000;**97**:3735-3740

[35] Liu J, Zhu J-K. A calcium sensor homolog required for plant salt tolerance. Science. 1998;**280**:1943-1945

[36] Shi H, Ishitani M, Kim C, Zhu J-K. The Arabidopsis thaliana salt tolerance gene SOS1 encodes a putative Na1/H1 antiporter. Proc Natl Acad Sci USA. 2000;**97**:6896-6901

[37] Attia H, Arnaud N, Karray N, Lachaâl M. Long-term effects of mild salt stress on growth, ion accumulation and superoxide dismutase expression on Arabidopsis rosette leaves. Physiologia Plantarum. 2008;**132**:293-305

[38] Chinnusamy V, Zhu J, Zhu J-K. Cold stress regulation of gene expression in plants. Trends in Plant Science. 2007;**12**(10):444-451

[39] Zhu T, Provart NJ. Transcriptional responses to low temperature and their regulation in Arabidopsis. Canadian Journal of Botany-Revue Canadienne de Botanique. 2003;**81**:1168-1174

[40] Lee B-H, Henderson DA, Zhu J-K. The Arabidopsis cold-responsive transcriptome and its regulation by ICE1. Plant Cell. 2005;**17**:3155-3175

[41] Liu Q, Kasuga M, Sakuma Y, Abe H, Miura S, Yamaguchi-Shinozaki K, et al. Two transcription factors, DREB1 and DREB2, with an EREBP/AP2 DNA binding domain separate two cellular signal transduction pathways in drought- and low-temperature-responsive gene expression, respectively, in Arabidopsis. Plant Cell. 1998;**10**:1391-1406

[42] Mittler R. Abiotic stress, the field environment and stress combination. Trends in Plant Science. 2006;**11**(1):15-19

[43] Shinwari ZK, Nakashima K, Miura S, et al. An Arabidopsis gene family encoding DRE/CRT binding proteins involved in low-temperature-responsive gene expression. Biochemical and Biophysical Research Communications. 1998;**250**:161-170

[44] Medina J, Bargues M, Terol J, Perez-Alonso M, Salinas J. The Arabidopsis CBF gene family is composed of three genes encoding AP2 domain-containing proteins whose expression is regulated by low temperature but not by abscisic acid or dehydration. Plant Physiology. 1999;**119**:463-469

[45] Dong CH, Agarwal M, Zhang Y, Xie Q, Zhu JK. The negative regulator of plant cold responses, HOS1, is a RING E3 ligase that mediates the ubiquitination and degradation of ICE1. Proc Natl Acad Sci USA. 2006;**103**:8281-8286

[46] Ulrich HD. Mutual interactions between the SUMO and ubiquitin systema: A plea of no contest. Trends in Cell Biology. 2005;**15**:525-532

Chapter 3

Plant Secondary Metabolites and Abiotic Stress Tolerance: Overview and Implications

Attiqa Rahman, Ghadeer M. Albadrani, Ejaz Ahmad Waraich, Tahir Hussain Awan, İlkay Yavaş and Saddam Hussain

Abstract

Plant secondary metabolites (PSM) are one of the major sources of industrially important products such as food additives due to their distinctive tastes, smells, and flavors. Unlike primary metabolites such as carbohydrates, lipids, and proteins, these secondary chemicals are not involved in plant growth, development, and reproduction but play a significant role in ecosystem functioning. These secondary biochemicals also play a key role in plant communication and defense, particularly under different environmental stresses. Plants may exhibit a defense response to combat these abiotic environmental stressors by generating a variety of PSMs to minimize cell and tissue damage. Secondary metabolites are very diverse (almost more than 200,000) in nature, majorly classified into terpenoids, phenolic compounds, nitrogen, and sulfur-containing secondary metabolites, separated based on biosynthetic pathways (shikimate pathway, mevalonic pathway, and tricarboxylic acid cycle pathway). This chapter summarizes the stimulating effects of different abiotic stressors (heavy metals, cold and high temperature, light, salinity, and drought) on secondary metabolite production. A major focus is given on the synthesis of secondary metabolite and accumulation in plants under stressful conditions, and their role in the regulation of plant defense.

Keywords: secondary metabolites, ecosystem regulator, abiotic stresses, phenolics, plant defense, terpenoids

1. Introduction

Primary and secondary metabolites are two categories of substances synthesized by plants. Primary metabolites that directly contribute to plant growth and development include lipids, proteins, and carbohydrates. On the other hand, secondary metabolites (SMs) are multipurpose metabolites that frequently participate in plant defense and environmental communications [1]. Interspecies communication, controlling the activity of enzymes, signaling, and defense are just a few of the tasks performed by SMs [2].

The SMs play a major role in the plant's capacity to endure abiotic stressors. The plant defense system has developed under a variety of stress detection systems, including transmembrane recognition, and the creation of SMs, which ultimately help plants to withstand these harsh or stressful conditions [3, 4].

Secondary metabolism also permits ecological interactions between plants and some other species and contributes to the capacity of plants to adjust and survive in response to environmental conditions throughout their lifecycle [5]. Plants are often at risk from a variety of abiotic stresses, such as oxidative and toxic metals, extreme cold, flooding, salinity, and drought [6]. Plants create a variety of low molecular weight chemicals known as secondary metabolites, such as anthocyanins, phenolic compounds, alkaloids, flavonoids, steroids, and terpenes. These phytochemicals are vital for a plant's protection, adaptation, and environmental adjustment [7].

Plant secondary metabolites are distinctive sources of flavorings, food additives, and medications with significant industrial use [8]. A transition from a rapid growth period to a stationary(inactive) phase in plants is typically due to the highest amounts of SMs production. Secondary metabolism is a crucial component of organic metabolism and biological processes; it mainly relies on primary metabolism for the contribution of the necessary enzymes, ATP, and cellular machinery and has a role to make a significant contribution to the producer organism's long-term survival. However, their manufactured organisms can grow and develop even without these secondary compounds, recommending that secondary metabolism is not necessary, at least for short-term or temporary survival [9].

Regulating plant tolerance at the cellular level is crucial for controlling stress responses and signal transduction systems [10]. For instance, the capacity of cotton to synthesize flavonoids and derivatives of cinnamic acid under drought stress tolerance indicated that they are highly effective at scavenging ROS [11], whereas the ability of reed plants to produce isoprene during heat-induced stress suggested that they have an effective antioxidant capacity that can quench oxygen [12]. There are more than 200,000 SMs known in plants [13]. Some SMs are unique to certain plant taxa, and their concentrations might differ between populations and individual plants depending on tissue type and specific plant development stage [14]. Although such SMs variations may be a result of genetic variability, their concentrations are also influenced by abiotic factors (growth circumstances) that are anticipated to get worse with climate change (such as heat stress, drought, and UV radiation) [15]. It has been believed that increased synthesis of the majority of SMs by a plant's mechanism for chemical defense response is linked to their resistance to stress and that this may be one explanation for the existence of variation in the type and quantity of SMs produced by different plant's taxonomic groups [16, 17].

This chapter elaborates on the sources and classification of plant secondary metabolites, their synthesis under unfavorable conditions, and their role in modulating plant growth and abiotic stress tolerance.

2. Sources and classification of secondary metabolites

A simple classification of plant secondary metabolites includes three main groups: terpenes (which include plant volatiles, cardiac glycosides, carotenoids, and sterols), phenolics (which include phenolic acids, coumarins, lignans, stilbenes, flavonoids, tannins, and lignin), and nitrogen-containing compounds (which include alkaloids and glucosinolates) [18].

2.1 Terpenoids

Terpenoids are the largest and most diverse collection of natural products and contain thousands of isoprene units ranging from hemiterpenes to rubber, varying in structure from linear to polycyclic molecules. Each terpene is produced by the condensed isoprene units, and all of them are classified based on the number of carbon (5C) units that make up the basic structure [19].

A monoterpene is a compound composed of two isoprene units (C10) and is found in a wide range of plants. Terpenoids were first reported in their volatile forms, where they contributed to abiotic stress tolerance and chemical communications between species [20]. Terpenes have several uses for attracting pollinators, protecting plants from herbivores [21], and also acting as deadly insecticides and insect repellents [22].

2.2 Phenolics

Phenolics are a crucial subclass of aromatic PSMs, with one or many acidic hydroxyl groups linked to phenyl-ring [23, 24]. Phenolics are a family of heterogeneous molecules made up of about 10,000 chemicals that are both soluble in water and organic solvents [25]. Plants, fungi, and bacteria can make these substances [26].

Phenolic compounds are classified based on their structural characteristics (phenols (C6), phenolic acids (C6-C1 or C6-C3), flavonoids (C6-C3-C6), and tannins (C6-C3-C6)). They can occasionally be further divided according to the degree of polymerization: (I) High molecular weight phenolic compounds, such as tannins, are water-soluble molecules that are strongly polymerized and fall under two subfamilies with distinct properties (II). The majority of saprotrophic organisms readily use low molecular weight phenolic compounds, which are discovered in most plants in glycosylated and solubilized form; however, some of them have specific biological activities such as cinnamic acids, in dicot angiosperms, the bulk of hydrolyzable tannins are present. Pro-anthocyanidins is another name for condensed tannins, which are polymers of flavan-3-ols (flavonoids) [27].

They serve significant roles in giving fruits, flowers, and seeds color, smell, and flavor, and they also have powerful antioxidant qualities because of their physical and chemical makeup [28]. Moreover, these substances have an impact on interactions between plants, such as allelopathic inhibition of targeted species growth [29].

2.3 Nitrogen-containing secondary metabolites

Glucosinolates, alkaloids, and cyanogenic glycosides make up a significant class of nitrogen-containing SMs. Alkaloids are basically alkaline compounds, which have low molecular weight and make up a large group of secondary metabolites [30]. The majority of alkaloids occur in their free state as N-oxides or mix with acids to form salts that can easily dissolve in water [24]. Alkaloids are identified due to the presence of at least one nitrogen atom, and they have a noticeable physiological impact on animal behavior; for instance, many of them have poisonous or analgesic effects (such as cocaine or morphine) [21].

Cyanogenic glucosides (CNglcs), also known as α-hydroxy nitrile glucosides, are naturally occurring bioactive plant secondary metabolites generated from amino acids and primarily composed of the sugar D-glucose [31]. According to Picmanová et al. [32], CNglcs are thought to be crucial for plant development, growth, and tolerance against different abiotic stressors [33].

2.4 Sulfur contains secondary metabolites

Phytoalexins, defensins, glutathione (GSH), and allicin are just a few of these substances that have been linked either directly or indirectly to plants' protection against microbial diseases [34]. The GSH is one of the primary forms of organic sulfur, which is present in soluble parts of plants and is essential for controlling plant development and growth as well as working as an antioxidant in cells under stress [35].

Higher plants synthesize glucosylates (GSL), a type of low molecular weight plant glucosides containing nitrogen and sulfur to boost their resistance against predators, competitors, and parasites. These compounds degrade into toxic or repellent volatile defense chemicals, for example, allylcys sulfoxides in allium and mustard oil and glucosides in *Cruciferae* [36]. A typical structural characteristic of Brassica phytoalexins is an indole or closely similar ring structure. These secondary metabolites are different compared to other well-known GSLs and appear to exclusively be produced by the plant family Cruciferae. Due to their high value, multiple research teams have looked at cruciferous phytoalexins as well as their biological activities over the past few decades [37].

3. Production sites and biosynthesis of secondary metabolites

Within plant cells, many secondary compounds are biosynthesized from primary metabolites, and their production can be stimulated under many abiotic elicitors and signaling molecules [38]. Under stressful environments, specific SMs conditionally accumulate in different plant sections [39]. The Krebs cycle and the Shikimate pathway are run where the precursors of metabolites are formed. The important starting materials for SMs are primary metabolites. Based on their chemical makeup, intended use, and distribution in plants, primary and SMs can be separated from one another. In most plants, the basic biosynthesis routes of metabolites remain conserved and most of the primary metabolites are present in all tissue types. Several basic metabolic frameworks have emerged because of the preservation of this metabolic core. Many modifications in fundamental structures result from regular glycosylation, methylation, hydroxylation, acylation, oxidation, phosphorylation, and prenylation, as well as from fewer chemical modifications brought on by the specialized enzymes.

The three main categories of SMs can be separated based on biosynthetic pathways. Phenolic substances are synthesized through the shikimate pathway. Terpenes are synthesized through the mevalonic pathway, while the N-containing substances are synthesized through tricarboxylic acid cycle pathway [40]. The precursor of the shikimate pathway (from the pentose phosphate pathway) is shikimic acid, which is produced when erythrose 4-phosphate and phosphoenolpyruvate from the glycolytic pathway combine. Phenylalanine, tyrosine, and tryptophan are the precursors for PSMs like phenolics, and N-containing compounds are produced *via* the shikimate pathway and also serve as the building blocks for protein synthesis [39]. In contrast to tryptophan (which is the precursor of alkaloids, phytoalexins, indole glucosinolates, and plant hormones like auxin), tyrosine further produces isoquinoline alkaloids, pigment betalains, and quinones (such as tocochromanols and plastoquinone) [41].

4. Metabolomic approaches for quantifying secondary metabolites

Understanding the kinds and quantities of SMs generated by the kingdom of plants is beneficial for plant research since this production reveals how plants have evolved in nature to meet various challenges [42]. We need to use a wide range of various analysis tools to get better handling of metabolites for plant metabolomics due to the wide variety of chemical classes and characteristics as well as their enormous dynamic range of metabolite concentrations in plants.

After the organic substrate has been extracted using a non-polar solvent, the only approach for terpene analysis is through GCMS (gas chromatography-mass spectroscopy). The principle behind the technique is the specific absorption of near-infrared electromagnetic radiation by different OCs (organic compounds), which is known as near-infrared spectrometry (NIRS). This method creates spectra reflecting the material's organic composition (nitrogen, lignin, cellulose, hemicelluloses, etc.) [43]. Another recently created comprehensive method is called "metabolomics," which enables quantitative determination and all metabolites' identifications in different tissue by using a variety of analytical methods such as gas/liquid chromatography (GC/LC) combined with mass spectrometry (MS) or nuclear magnetic resonance (NMR) spectroscopy. Metabolomics provides a full snapshot of the chemical makeup of a tissue (such as a leaf) at a specific moment, enabling somebody to focus on the least understood chemicals and evaluate or recognize metabolites that have insufficient information [44].

Folin-Ciocalteu method used to calculate the concentration of OH^- phenolic compounds that are bound to the benzene ring is the method most frequently employed to examine the phenolic fraction of plants [45]. The Folin-Ciocalteu reagent reacts with phenolic groups to produce a blue complex, which is the basis for this procedure [46].

The STME (Silicone tubing microextraction) is a technique created by Mohney et al. [47] and is one of the newer techniques that open up new opportunities for studying secondary metabolites in soils. Direct placement of sorbent microtubes in the soil allows for *in situ* monitoring of allelochemicals released from roots [48].

5. Secondary metabolite accumulation under various environmental factors

The production and accumulation of PSMs in tissues are tightly regulated in a spatiotemporal way and influenced by many abiotic factors [49]. Environmental factors affect how PSMs produce and accumulate in different plants [50]. The alkaloid content of *Catharanthus roseus* seedlings under salt stress and water stress (drought) was significantly higher than when it is under control conditions [51]. Different abiotic stress conditions also have a substantial impact on the formation of phenolic chemicals [52]. *Oryza sativa* stimulates phenolics secretions in the stele and epidermis of roots in alkaline conditions, which effectively boosts ion absorption and reduces iron-deficiency reactions [53].

Polyphenols that are also called flavonoids are antioxidants and necessary for plant tolerance against different abiotic stresses [54]. In model plants, excessive accumulation of various flavonoids such as kaempferol, quercetin, and cyanidin is well known [55]. Heat and salt stress increased flavonoid accumulation in *O. sativa*, which enhanced tolerance in rice against these stresses [56]. Short-wavelength radiation causes a number of flavanol glycosides including quercetin and kaempferol glycosides, which strengthen plant defenses against different stresses [57]. In plants of Brassicaceae family, glucosinolates are significant precursors to several active components [58]. Strong light, high temperatures, and drought caused more accumulation of glucosinolates in *Brassica rapa* [59]. *Brassica oleracea* show strong tolerance against chilling and freezing, and it is suggested that the defensive mechanism enabling this tolerance involves the glucosinolates concentrations induced by low temperature [60].

6. Plant secondary metabolites and abiotic stress tolerance

Metabolic processes that lead to the accumulation of natural products are affected by the concentrations of various PSMs, which greatly affect growth conditions. Numerous typical reactions occur when abiotic stresses affect the stimulation of PSMs.

6.1 Heavy metals

Zinc, manganese, nickel, and iron are essential for the development of photosystems (I & II) and different enzymes in plant cells [61]. An excess of various metals especially toxic metals is harmful to plants; as a result, plant cells have systems in place to prevent these metals' poisonous buildup. Recent research has focused on the development of SMs within plants under heavy metal stress [62]. The formation of different photosynthetic pigments, sugars, proteins, and non-protein thiols is affected by heavy metals at physiological and metabolic levels in plants. By altering certain parts of secondary metabolism, metals can change how bioactive molecules are produced [63].

Secondary metabolite synthesis is also regulated by metal ions (europium, silver, lanthanum, and cadmium) and oxalates [64]. Urease enzyme, which is an essential component of the trace metal nickel (Ni), is required for the development of plants [65]. It has been demonstrated that Cu^{2+} and Cd^{2+} increase the yields of secondary metabolites such as shikonin [66]. Babula and colleagues [67] examined the physiological reactions of *Hypericum perforatum* plants to cadmium stress in several tissues, specifically in the shoots and the roots. Their findings revealed an increase in phenolic acids (ferulic acid), and on the other hand, there was a decrease in flavonoids (epicatechin and procyanidin) in shoots as well as in roots. It is interesting to note that PAL (phenylalanine ammonia lyase), the first gene to intervene in the phenylpropanoid pathway, was found to be directly correlated with heavy metal accumulation [68]. This showed how heavy metals affect the genes that are responsible for producing phenylpropanoids and explained why phenolic acids build up in heavy metal-stressed plant cells. To conserve energy plants, produce phenolic acids (hydroxycinnamic acids), and may prefer to invest in the first steps of the pathway rather than activating the genes that would otherwise interfere with the proceeding steps (which result in the synthesis of flavonoids and anthocyanins). It was previously mentioned that phenolic compounds are produced as a defensive mechanism in response to toxic metal stress. Phenolics are powerful Cd chelators and the roots of *Matricaria chamomilla* produce more of these compounds than other plants [69].

In plant cells, oxidants and antioxidants coexist in a dynamic balance that prevents ROS buildup [70]. Secondary metabolites play a well-established role in reducing ROS stress [71]. The plant secondary metabolites that can combat ROS and prevent oxidative stress are polyphenols and terpenes [72]. Their scavenging abilities are also caused by these molecules, it is important to note that the antioxidant characteristic of flavonoids is determined by their OH^- groups that provide electrons and hydrogen to radicals to stabilize them [73]. A $CdSO_4$ treatment led to an increase in protective soluble phenolic compounds within the woody species Populus x canescens [74]. In contrast to the wood, where ROS were created at a faster rate these chemicals were more prevalent in the bark. These findings showed that different organs of the same tree exhibit diverse responses to heavy metals and that these responses are correlated with the capability of plants to produce different SMs [75].

6.2 Temperatures (cold and high temperatures)

The temperature has a significant impact on plant ontology and metabolic activity, and extreme heat can expedite the senescence of leaves. Thermal treatments were observed to marginally reduce carotenoids in Brassicaceae, including β-carotene [76]. Temperatures and the phenological stage had an impact on the production of SMs in *Rhodiola rosea* clones [77], and increased levels of toxic metals boosted SMs production with a synergistic action associated [78]. Within suitable temperature ranges, plants can grow and develop more effectively. The development and production of plants may be negatively impacted by low and high temperatures [79]. Heat stress affects plants that are growing in hot environments. Stomatal conductance and net CO_2 fixation drop due to heat stress are linked to decreased plant growth and yield. Heat stress in plants and SMs biosynthesis are related to one another [80]. A decrease in the photochemical efficiency of photosystem II is seen in plants developing under heat stress. A review of the literature found that plants under heat stress often produce more SMs, but some studies also showed a decrease in SMs production under elevated temperatures; ginsenoside levels were increased in *Panax quinquefolius* plants that were cultivated under elevated temperature stress [65].

Among the most detrimental abiotic stressors affecting temperate plants is low-temperature stress. Due to seasonal temperature changes, several species' metabolisms have modified in the fall to contain more of a variety of cryo-protective compounds for enhancing their capacity to survive cold temperatures [81]. When a temperate plant overwinters, its metabolism is switched to the synthesis of molecules that act as cryoprotectants, such as sugar alcohols, and low molecular weight nitrogenous compounds [81]. Low-temperature stress inhibits metabolic processes, water absorption, and cellular dehydration in many plants [82]. Freezing temperatures caused photosynthesis in *Capsicum annuum* plants ultimately reducing the plant growth. Cold acclimation occurs when plants that are growing at low temperatures show significant changes in a variety of physiochemical and molecular mechanisms allowing plants to withstand these cold temperature stresses. Moreover, information about the decline in photosynthetic pigments and total soluble protein content in plants during cold temperatures has been documented in the literature [83]. The production and storage of SMs were noticeably decreased under low-temperature stress [80]. Phenolic synthesis is also observed to be increased by cold stress [84]. The relationship between temperature and the production of alkaloids has been observed, particularly with high temperatures being preferred to trigger alkaloid production. At low temperatures, the accumulation of the alkaloids was constrained in dry *Papaver somniferum* [85]. In contrast to the control, the overall phenolic acid content and isoflavonoids (genistein, daidzein, and genistin) in soybean (*Glycine max*) roots increased when these roots were treated at a cold temperature for 24 hours with genistin showing the greatest rise of 310% [86]. Christie et al. [87]

documented the development of anthocyanins during low-temperature stress. *Pinus pinaster* undergoes modifications in its endogenous jasmonates because of cold and water stressors [88].

According to Lei et al. [89], melatonin protects carrot suspension cells against cold-induced apoptosis *via* upregulating the polyamines (putrescine and spermine). According to a recent study by Zhao et al. [90], melatonin has been shown to increase the longevity of *Rhodiola crenulata* cryopreserved callus. Kovacs et al. [91] observed that when wheat (*Triticum aestivum* L.) leaves are subjected to low temperatures putrescine accumulates (6–9 times), spermidine accumulates less, and spermine declines little. Moreover, under low-temperature stress, alfalfa (*Medicago sativa* L.) also accumulates putrescine [92]. According to Hummel et al. [93], agmatine and putrescine levels have been linked to enhanced levels of cold tolerance, and they may serve as a useful indicator of this trait in *P. antiscorbutica* seedlings. *Perilla frutescens* suspension cultures showed a striking reduction in anthocyanin production at an elevated temperature of 28°C, which was the greatest at 25°C [94]. Similar findings on anthocyanin productivity at its maximum production level in *Daucus carota* cell suspension cultures were described [95]. Under the influence of various temperatures, *Beta vulgaris* hairy root cultures were observed and examined for a release of these pigments [96]. The ideal temperature ranges for each plant species and cultivar are unique, and any variation from those limits may have an impact on biomass and the production of SMs.

6.3 Salinity stress

Salinity stress affects plant growth and the production of bioactive compounds [97, 98]. Many plant species create phenolics to defend themselves against different abiotic stress conditions, including salinity, and their buildup is correlated with plant species' antioxidant capacities [99]. Proline levels in the roots of salt-tolerant alfalfa plants increased quickly according to research by Petrusa and Winicov, the increase was gradual in salt-sensitive plants [100]. Many plants have also observed an increase in polyphenol content in various tissues under salt stress [101]. Navarro et al. [102] found that red peppers had an enhanced total phenolic content at a moderate salinity level. It has been demonstrated that plant polyamines influence how plants react to salinity. There have been reports of alterations in polyamine levels caused by salinity in *Helianthus annuus* L. (sunflower) roots [103]. The effects of KCl treatment on amounts of total phenolics and flavonoids in *C. cardunculus* and Cardunculus var. altilis leaves were more pronounced than those of the other two chloride salts (NaCl and $CaCl_2$) [104] (**Table 1**).

6.4 Drought stress

Drought stress is one of the major abiotic stresses that affect plant development and growth [93, 113]. Drought disrupts cellular homeostasis by affecting proteins, carbohydrates, lipids, and DNA. It has an impact on the plant's height, root growth, and leaf area (LA) [114, 115]. Moreover, drought has a significant impact on the physiology of plants, including osmotic potential, stomatal conductance, rate of photosynthesis, pressure potential, and transpiration rates [116]. Drought stress poses a serious threat to sustainable agriculture since it has a negative impact on crop yield globally. However, in response to drought stress, plants have evolved several morphological, physiological, biochemical, and phonological mechanisms [117].

Willow (Salix) leaves were shown to contain more flavonoids and phenolic acids during a drought, which frequently results in oxidative stress [118]. Changes in the

Plant species	Salinity levels	Effects on the concentrations of SMs	References
Solanum nigrum	0, 50, 100, & 150 mM	Salinity stress increased the expression of the flavonoid genes, which in turn increased the synthesis of quercetin 3-d-glucoside and lutein. Moreover, certain carotenoid-related genes, such as phytoene synthase 2 and lycopene cyclase were just overexpressed in response to salt stress.	[105]
Zea Mays	0 & 100 mM	Salinity stress increased the total phenolic and flavonoid levels.	[106]
O. sativa	0 & 25 mM	Grain nutritional quality, including antioxidant activities, anthocyanins, and total phenolics, was significantly improved under salinity stress.	[107]
Triticum astivum	0 & 150 mM	Salinity stress increased phenolics, particularly at the booting stage.	[108]
Lepidium sativum	30, 60, 90, & 120 mM	Accumulation of flavonoids and phenolic compounds in *L. sativum* was increased under salinity stress.	[109]
Saccharum officinarum	12.5 & 6.8 dS/m	Salinity stress altered the production of flavones, anthocyanins, and soluble phenolics in two sugarcane clones, CP-4333 and HSF-240.	[110]
Fagopyrum esculentum	0.50,100, & 200 mM	Carotenoids, phenolics, and antioxidant activity changed noticeably when buckwheat plants were exposed to different salinity levels. Comparing plants grown under non-saline conditions to plants growing under various salinity doses, it is evident that the concentration of phenolic chemicals increased significantly.	[111]
Salvia mirzayanii	4.5, 6.8, & 9.1 dS/m	*S. mirzayanii's* antioxidant activity, phenolic content, and volatile compounds (Bicyclogermacrene, 1,8-cineole, and -terpinyl acetate) were increased under salinity stress.	[112]

Table 1.
Influence of salinity stress on the biosynthesis and accumulation of different PSMs.

ratio of chlorophylls "*a*" & "*b*" and carotenoids were affected by the drought [119]. Cotton under drought stress was shown to have less chlorophyll [120], as was in *C. roseus* [121]. In *Chenopodium quinoa,* drought circumstances reduced the amount of saponins from 0.46% dry weight (dw) in plants growing in low water deficit settings to 0.38% in plants growing in high water deficit situations [121]. A number of SMs generated by plants are beneficial for fostering drought resistance [96, 122].

A different study found that applying drought stress improved the quality of significant SMs in *Artemisia annua* [80]. Similarly, *Glechoma longituba* grown in drought conditions showed an increase in total flavonoids [123]. Significant changes were seen in the contents of several macronutrients, proline, carbohydrates, and essential oils in *Ocimum americanum* and *Ocimum basilicum* under water-limited circumstances [124] (**Table 2**).

Plant species	Drought levels	Effects on the concentrations of SMs	References
Achillea filipendulina	25, 50,75, &100% (field capacity)	Drought increased total phenolic and flavonoid concentrations.	[125]
Zea mays	PEG-induced (0.6 MPA)	Reduced phenolic compounds and decreased plant biomass occurred under drought.	[126]
O. sativa	25, 45, 65, & 85% Soil moisture	Drought increased the production of flavonoids, phenolics particularly in tolerant genotypes.	[127]
G. max	Control (−15 to −20) Drought (−90 & -100KPA)	Total phenolics and lignin levels significantly increased under drought	[128]
Carthamus tinctorius	25 and 50% water deficit	Plants under mild water scarcity produced more phenolic compounds, whereas plants under severe drought showed a clear drop in phenolics. Likewise, plants exposed to moderate drought had much higher levels of carotenoids.	[129]
Vitis vinifera	30 & 70% Soil moisture	In plants under drought stress, the amounts of the phenolic compounds significantly dropped.	[130]

Table 2.
Drought-induced alterations in the biosynthesis and storage of PSMs.

6.5 Light

The physiological reactions of various plant species and even cultivars to exposure to light conditions, such as photoperiod or small durations connected to the generation of SMs [131]. Light is a physical element that is widely established to have an impact on metabolite synthesis. In *Z. officinale* callus cultivation, light can enhance the formation of such secondary metabolites as gingerol and zingiberene [5]. Hence, the amounts of phenolics have been found to increase in direct proportion to light intensity.

Due to shorter light duration, many plant portions have significantly lower endogenous levels of coumarins. Furthermore, the prolonged period of light markedly enhanced the number of coumarins [132]. American ginseng (*P. quinquefolius*) plants that were exposed to direct sunlight for a longer period produced more ginsenoside in their roots than those that were exposed for a shorter time [133].

Blue light was found to have the greatest impact on SMs in *Scutellaria laterifora* shoot cultures, and their connection with PGRs (Plant growth regulators) was found [134]. The effects of various light spectra from light-emitting diode sources on the production of SMs were seen when *Peucedanum japonicum* callus cultures were exposed to them. The red and blue light was shown to be the most effective [135]. Based on the length of the cell suspension cultures of *Artemisia absinthium*, light and dark incubation conditions had a significant impact on the generation of biomass [136]. Different species have different effects on how light affects plant growth and development [131].

According to Liang et al. [137], UV-B radiation may cause a decrease in chlorophyll content while increasing flavonoid content and PAL activity. Root flavonoids in

Pisum sativum plants were elevated by UV light (300–400 nm) [138]. Recent research showed that photoperiod regimes influence endogenous indoleamines (serotonin and melatonin) in farmed green algae *Dunaliella bardawil* [139]. In primary and secondary metabolism as well as a number of plant developmental processes, light is widely known to be essential [140]. Several studies have revealed that light sources directly induced the synthesis of crucial secondary metabolites, such as anthocyanins, artemisinin, caffeic acid derivatives, and flavonoids [141]. Regvar et al. [142] compared the effects of UV irradiation on different concentrations of rutin, catechin, and quercetin in *Fagopyrum esculentum* and *F. tataricum*, and they discovered that *F. esculentum* was found to have more quercetin when exposed to the elevated UV irradiation. Markham et al. [143] investigated the C-glycosyl flavone content of various rice cultivars under UV-B light and discovered that C-glycosyl flavones were enriched in a UV-tolerant rice cultivar but lacking in a sensitive cultivar.

7. Conclusions

This chapter explains the importance of secondary metabolites in plants' defense against abiotic stresses such as heavy metals, flooding, salinity, and drought. These metabolites are produced in response to environmental stressors and are regulated depending upon growth circumstances and developmental stage. There are three main groups of secondary metabolites: terpenoids, phenolics, and nitrogen-containing compounds. Higher plants synthesize GSL (N & S containing secondary metabolites) to boost their resistance against predators, competitors, and parasites. The biosynthetic pathways of these SMs are distinct and use different precursors, with the shikimate pathway producing phenolic substances, the mevalonic pathway producing terpenes, and the tricarboxylic acid cycle pathway producing nitrogen-containing compounds. Understanding the types and quantities of secondary metabolites in plants is important for plant research, as it reveals how plants have evolved to cope with various challenges. Metabolomics is a comprehensive method used to identify and quantify all metabolites in different tissues. Flavonoids and glucosinolates are two examples of secondary metabolites that are important for plant tolerance against different abiotic stresses. Plant breeders have the potential to develop new plant varieties with increased tolerance to various abiotic stresses by selectively incorporating specific secondary metabolites. In the context of climate change, where plants will face more extreme environmental conditions, this could be particularly valuable.

Acknowledgements

The authors acknowledge the support from Princess Nourah bint Abdulrahman University Researchers Supporting Project number (PNURSP-HC2022/4), Princess Nourah bint Abdulrahman University, Riyadh, Saudi Arabia.

Conflict of interest

The authors declare that there is no conflict of interest.

Author details

Attiqa Rahman[1], Ghadeer M. Albadrani[2], Ejaz Ahmad Waraich[3], Tahir Hussain Awan[4], İlkay Yavaş[5] and Saddam Hussain[2,3]*

1 Department of Botany, University of Agriculture, Faisalabad, Pakistan

2 Department of Biology, Princess Nourah bint Abdulrahman University, Riyadh, Saudi Arabia

3 Department of Agronomy, University of Agriculture, Faisalabad, Pakistan

4 Rice Research Institute, Kala Shah Kaku, Pakistan

5 Department of Plant and Animal Production, Adnan Menderes University, Aydın, Turkey

*Address all correspondence to: sadamhussainuaf@gmail.com

References

[1] Jan R, Asaf S, Numan M, Kim KM. Plant secondary metabolite biosynthesis and transcriptional regulation in response to biotic and abiotic stress conditions. Agronomy. 2021;**11**(5):968

[2] Hussain MS, Fareed S, Ansari S, Rahman MA, Ahmad IZ, Saeed M. Current approaches toward production of secondary plant metabolites. Journal of Pharmacy & Bioallied sciences. 2012;**4**(1):10

[3] Dawid C, Hille K. Functional metabolomics—A useful tool to characterize stress-induced metabolome alterations opening new avenues towards tailoring food crop quality. Agronomy. 2018;**8**(8):138. DOI: 10.3390/agronomy8080138

[4] Ashraf MA, Iqbal M, Rasheed R, Hussain I, Riaz M, Arif MS. Environmental stress and secondary metabolites in plants: An overview. Plant Metabolites and Regulation Under Environmental Stress. 2018:153-167. DOI: 10.1016/B978-0-12-812689-9.00008-X

[5] Akula R, Ravishankar GA. Influence of abiotic stress signals on secondary metabolites in plants. Plant Signaling and Behavior. 2011;**6**(11):1720-1731

[6] Flora SJS, Shrivastava R, Saito K. Phytochemical genomics—A new trend. Current Opinion in Plant Biology. 2013;**16**(3):373-380. DOI: 10.1016/j.pbi.2013.04.001

[7] Wink M. Evolution of secondary metabolites from ecological and molecular phenological perspective. Phytochemistry. 2013;**64**:3-19

[8] Ravishankar GA, Rao SR. Biotechnological production of phytopharmaceuticals. Journal of Biochemistry, Molecular Biology, and Biophysics. 2000;**4**:73-102

[9] Roze LV, Chanda A, Linz JE. Compartmentalization and molecular traffic in secondary metabolism: A new understanding of established cellular processes. Fungal Genetics and Biology. 2011;**48**:35-48

[10] Fraire-Velázquez S, Balderas-Hernández VE. Abiotic stress in plants and metabolic responses. In: Vahdati K, Leslie C, editors. Abiotic Stress-Plant Responses and Applications in Agriculture. New York: London, UKInTech Open Science; 2013. pp. 25-48

[11] Yildiz-Aktas L, Dagnon S, Gurel A, Gesheva E, Edreva A. Drought tolerance in cotton: Involvement of non-enzymatic ROS-scavenging compounds. Journal of Agronomy and Crop Science. 2009;**195**(4):247-253

[12] Velikova V, Pinelli P, Loreto F. Consequences of inhibition of isoprene synthesis in Phragmites australis leaves exposed to elevated temperatures. Agriculture, Ecosystems and Environment. 2005;**106**:209-217

[13] Neilson EH, Goodger JQD, Woodrow IE, Møller BL. Plant chemical defense: At what cost? Trends in Plant Science. 2013;**18**:250-258

[14] Ritmejeryte E, Boughton BA, Bayly MJ, Miller RE. Unique and highly specific cyanogenic glycoside localization in stigmatic cells and pollen in the genus Lomatia (Proteaceae). Annals of Botany. 2020;**126**:387-400

[15] Sampaio BL, Edrada-Ebel R, Da Costa FB. Effect of the environment

on the secondary metabolic profile of Tithonia diversifolia: A model for environmental metabolomics of plants. Scientific Reports. 2016;**6**:1-11

[16] Goyal S, Lambert C, Cluzet S, Mérillon JM, Ramawat KG. Secondary metabolites and plant defense. In: Mérillon JM, Ramawat KG, editors. Plant defense: biological control. Netherlands: Springer; 2012. pp. 109-138

[17] Ngo LT, Okogun JI, Folk WR. 21st-century natural product research and drug development and traditional medicines. Natural Product Reports. 2013;**30**:584-592

[18] Bizzo HR, Silveira D, Gimenes MA, Tânia da S, Agostini-Costa RF, Vieira HR, et al. Gimenes. Chromatography and its Applications. 2012:131

[19] Mahmoud SS, Croteau RB. Strategies for transgenic manipulation of monoterpene biosynthesis in plants. Trends in Plant Science. 2002;7:366-373

[20] Langenheim JH. Higher-plant terpenoids – A phytocentric overview of their ecological roles. Journal of Chemical Ecology. 1994;**20**:1223-1280

[21] Hopkins WG. Physiologie Vegetale. Bruxelles, Belgium: De Boeck & Larcier; 2003

[22] Malcovska SM, Ducaiova Z, Maslaáková I, Backor M. Effect of silicon on growth, photosynthesis, oxidative status and phenolic compounds of maize (*Zea mays* L.) grown in cadmium excess. Water, Air Soil Pollut. 2014;**225**(8):1

[23] Ali M, Abbasi BH, Ihsan-ul-haq. Production of commercially important secondary metabolites and antioxidant activity in cell suspension cultures of *Artemisia absinthium* L. Industrial Crops and Products. 2013;**49**:400-406

[24] Evans WC. Alkaloids. In: Evans WC, editor. Trease and Evans Pharmacognosy. Edinburgh, UK: Elsevier; 2009. pp. 353-415

[25] Taiz L, Zeiger E. Secondary metabolites, and plant defense. Plant Physiology. 2006;**13**:125

[26] Hättenschwiler S, Vitousek P. The role of polyphenols in terrestrial ecosystem nutrient cycling. Trends in Ecology & Evolution. 2000;**15**:238-243

[27] Moreira LR, Milanezi NV, Filho EX. Enzymology of plant cell wall breakdown: an update. Routes to cellulosic ethanol. New York, NY: Springer; 2010;73-96

[28] Mierziak J, Kostyn K, Kulma A. Flavonoids as important molecules of plant interactions with the environment. Molecules. 2014;**19**:16240-16265

[29] Weston LA, Mathesius U. Flavonoids: Their structure, biosynthesis, and role in the rhizosphere, including allelopathy. Journal of Chemical Ecology. 2013;**39**:283-297

[30] Khan S, ur Rahman L. Pathway modulation of medicinal and aromatic plants through metabolic engineering using agrobacterium tumefaciens. In: Jha S, editor. Transgenesis and Secondary Metabolism. Cham: Springer International Publishing; 2017. pp. 431-462

[31] Møller BL. Functional diversifications of cyanogenic glucosides. Current Opinion in Plant Biology. 2010;**13**:338-347. DOI: 10.1016/j.pbi.2010.01.009

[32] Pičmanová M, Neilson EH, Motawia MS, Olsen CE, Agerbirk N, Gray CJ, et al. A recycling pathway for cyanogenic glycosides is evidenced by

the comparative metabolic profiling in three cyanogenic plant species. The Biochemical Journal. 2015;**469**:375-389

[33] Hayes CM, Weers BD, Thakran M, Burow G, Xin Z, Emendack Y, et al. Discovery of a dhurrin QTL in sorghum: co-localization of dhurrin biosynthesis and a novel stay green QTL. Crop Science. 2016;**56**:104-112. DOI: 10.2135/cropsci2015.06.0379

[34] Grubb C, Abel S. Glucosinolate metabolism and its control. Trends in Plant Science. 2006;**11**:89-100

[35] Kang SY, Kim YC. Decursinol and decursin protect primary cultured rat cortical cells from glutamate-induced neurotoxicity. The Journal of Pharmacy and Pharmacology. 2007;**59**(6):863-870

[36] Leustek T. Sulfate metabolism. In: Somerville CR, Meyerowitz EM, editors. The Arabidopsis Book. Rockville, MD: American Society of Plant Biologists; 2002. DOI: 10.1199/tab.0009

[37] Pagare S, Bhatia M, Tripathi N, Pagare S, Bansal YK. Secondary metabolites of plants and their role: Overview. Current Trends in Biotechnology and Pharmacy. 2015;**9**(3):293-304

[38] Narayani M, Srivastava S. Elicitation: A stimulation of stress in vitro plant cell/tissue cultures for enhancement of secondary metabolite production. Phytochemistry Reviews. 2018;**16**(6):1227-1252

[39] Ahuja I, Kissen R, Bones AM. Phytoalexins in defense against pathogens. Trends in Plant Science. 2012;**17**:73-90

[40] Jamwal K, Bhattacharya S, Puri S. Plant growth regulator mediated consequences of secondary metabolites in medicinal plants. Journal of Applied Research on Medicinal and Aromatic Plants. 2018;**9**:26-38

[41] Maeda H, Dudareva N. The shikimate pathway and aromatic amino acid biosynthesis in plants. Annual Review of Plant Biology. 2012;**63**:73-105

[42] Tohge T, Watanabe M, Hoefgen R, Fernie AR. The evolution of phenylpropanoid metabolism in the green lineage. Critical Reviews in Biochemistry and Molecular Biology. 2013;**48**:123-152

[43] Birth GS, Hecht HG. The physics of near-infrared reflectance. In: Williams P, Norris K, editors. Near-infrared Technology in the Agricultural and Food Industries. St. Paul, MN, USA: American Association of Cereal Chemists; 1987. pp. 1-16

[44] Breitling R, Ceniceros A, Jankevics A, Takano E. Metabolomics for secondary metabolite research. Metabolites. 2013;**3**:1076-1083

[45] Bärlocher F, MAS G. Total phenolics. In: Graca MAS, Bärlocher F, Gessner MO, editors. Methods to Study Litter Decomposition: A Practical Guide. New York, NY, USA: Springer; 2005. pp. 97-110

[46] Everette JD, Bryant QM, Green AM, Abbey YA, Wangila GW, Walker RB. A thorough study of reactivity of various compound classes toward the Folin–Ciocalteu reagent. Journal of Agricultural and Food Chemistry. 2010;**58**:8139-8144

[47] Mohney BK, Matz T, Lamoreaux J, Wilcox DS, Gimsing AL, Mayer P, et al. In situ silicone tube microextraction: A new method for undisturbed sampling of root-exuded thiophenes from marigold (*Tagetes erecta* L.) in soil. Journal of Chemical Ecology. 2009;**35**:1279-1287

[48] Weidenhamer JD, Mohney BK, Shihada N, Rupasinghe M. Spatial and temporal dynamics of root exudation: How important is heterogeneity in allelopathic interactions? Journal of Chemical Ecology. 2014;**40**:940-952

[49] Yang CQ, Fang X, Wu XM, Mao YB, Wang LJ, Chen XY. Transcriptional regulation of plant secondary metabolism. Journal of Integrative Plant Biology. 2012;**54**:703-712. DOI: 10.1111/j.1744-7909.2012.01161. x

[50] Gupta ML, Dutta A. Stress-mediated adaptive response leading to genetic diversity and instability in metabolite contents of high medicinal value: An overview on Podophyllum hexandrum. OMICS. 2011;**15**:873-882. DOI: 10.1089/omi.2011.0096

[51] Hassan FAS, Ali E, Gaber A, Fetouh MI, Mazrou R. Chitosan nanoparticles effectively combat salinity stress by enhancing antioxidant activity and alkaloid biosynthesis in *Catharanthus roseus* (L.) G Don. Plant Physiology and Biochemistry. 2021;**162**:291-300. DOI: 10.1016/j.plaphy.2021.03.004

[52] Sharma A, Shahzad B, Rehman A, Bhardwaj R, Landi M, Zheng B. Response of the phenylpropanoid pathway and the role of polyphenols in plants under abiotic stress. Molecules. 2019;**24**:2452. DOI: 10.3390/molecules24132452

[53] Li H, Chen H, Deng S, Cai H, Shi L, Xu F, et al. Inhibition of nitric oxide production under alkaline conditions regulates iron homeostasis in rice. Physiologia Plantarum. 2021b;**172**:1465-1476. DOI: 10.1111/ppl.13333

[54] Laoue J, Fernandez C, Ormeno E. Plant flavonoids in Mediterranean species: A focus on flavonols as protective metabolites under climate stress. Plants. 2022;**11**:172. DOI: 10.3390/plants11020172

[55] Nakabayashi R, Mori T, Saito K. Alternation of flavonoid accumulation under drought stress in Arabidopsis thaliana. Plant Signaling & Behavior. 2014;**9**:e29518. DOI: 10.4161/psb.29518

[56] Jan R, Kim N, Lee SH, Khan MA, Asaf S, Lubna, et al. Enhanced flavonoid accumulation reduces combined salt and heat stress through the regulation of transcriptional and hormonal mechanisms. Frontiers in Plant Science. 2021;**12**:796956. DOI: 10.3389/fpls.2021.796956

[57] Rechner O, Neugart S, Schreiner M, Wu S, Poehling HM. Can narrow-bandwidth light from UV-A to green alter secondary plant metabolism and increase *Brassica* plant defenses against aphids? PLoS One. 2017;**12**:e0188522. DOI: 10.1371/journal.pone.0188522

[58] Moreno DA, Carvajal M, Lopez-Berenguer C, Garcia-Viguera C. Chemical and biological characterization of nutraceutical compounds of broccoli. Journal of Pharmaceutical and Biomedical Analysis. 2006;**41**:1508-1522. DOI: 10.1016/j.jpba.2006.04.003

[59] Rao SQ, Chen XQ, Wang KH, Zhu ZJ, Yang J, Zhu B. Effect of short-term high temperature on the accumulation of glucosinolates in *Brassica rapa*. Plant Physiology and Biochemistry. 2021;**161**:222-233. DOI: 10.1016/j.plaphy.2021.02.013

[60] Ljubej V, Radojci Redovnikovic I, Salopek-Sondi B, Smolko A, Roje S, Samec D. Chilling and freezing temperature stress differently influence glucosinolates content in *Brassica oleracea* var. acephala. Plants. 2021;**10**:1305. DOI: 10.3390/plants10071305

[61] Page V, Feller U. Heavy metals in crop plants: Transport and redistribution processes on the whole plant level. Agronomy. 2015;**5**:447-463. DOI: 10.3390/agronomy5030447

[62] Eghbaliferiz S, Iranshahi M. Prooxidant activity of polyphenols, flavonoids, anthocyanins, and carotenoids: An updated review of mechanisms and catalyzing metals. Phytotherapy Research : PTR. 2006;**30**:1379-1139

[63] Verpoorte R, Contin A, Memelink J. Biotechnology for the production of plant secondary metabolites. Phytochemistry Reviews. 2002;**1**:13-25

[64] Marschner H. Mineral Nutrition of Higher Plants. London: Academic Press; 1995. p. 889

[65] Jochum GM, Mudge KW, Thomas RB. Elevated temperatures increase leaf senescence and root secondary metabolite concentrations in the understory herb *Panax quinquefolius* (Araliaceae). American Journal of Botany. 2007;**94**(5):819-826. DOI: 10.3732/ajb.94.5.819

[66] Ohlsson AB, Berglund T. Effect of high MnSO4 levels on ardenolide accumulation by Digitalis lanata tissue cultures in light and darkness. Journal of Plant Physiology. 1989;**135**:505-507

[67] Babula P, Klejdus B, Kovacik J, Hedbavny J, Hlavna M. Lanthanum rather than cadmium induces oxidative stress and metabolite changes in *Hypericum perforatum*. Journal of Hazardous Materials. 2015;**286**:334-342. DOI: 10.1016/j.jhazmat.2014.12.060

[68] Ma C, Liu H, Guo H, Musante C, Coskun SH, Nelson BC, et al. Defense mechanisms and nutrient displacement in Arabidopsis thaliana upon exposure to CeO2 and In2O3 nanoparticles. Environmental Science. Nano. 2016;**3**:1369-1379. DOI: 10.1039/C6EN00189K

[69] Kovácik J, Klejdus B. Dynamics of phenolic acids and lignin accumulation in metal-treated *Matricaria chamomilla* roots. Plant Cell Reports. 2008;**27**:605-661

[70] Lobo V, Patil A, Phatak A, Chandra N. Free radicals, antioxidants, and functional foods: Impact on human health. Pharmacognosy Reviews. 2010;**4**:118-112

[71] Bartwal A, Mall R, Lohani P, Guru SK, Arora S. Role of secondary metabolites and brassinosteroids in plant defense against environmental stresses. Journal of Plant Growth Regulation. 2013;**32**(1):216-232

[72] Sharma P, Jha AB, Dubey RS, Pessarakli M. Reactive oxygen species, oxidative damage, and antioxidative defense mechanism in plants under stressful conditions. Journal of Botany. 2012;**9**(8):681. DOI: 10.1155/2012/217037

[73] Heim KE, Tagliaferro AR, Bobilya DJ. Flavonoid antioxidants: Chemistry, metabolism and structure-activity relationships. The Journal of Nutritional Biochemistry. 2002;**13**:572-584

[74] He J, Qin J, Long L, Ma Y, Li H, Li K, et al. Net cadmium flux and accumulation reveal tissue-specific oxidative stress and detoxification in Populus × canescens. Physiologia Plantarum. 2011;**143**:50-63

[75] Berni R, Luyckx M, Xu X, Legay S, Sergeant K, Hausman JF, et al. Reactive oxygen species and heavy metal stress in plants: Impact on the cell wall and secondary metabolism. Environmental

and Experimental Botany. 2019;**161**:98-106

[76] Morison JIL, Lawlor DW. Interactions between increasing CO2 concentration and temperature on plant growth. Plant, Cell & Environment. 1999;**22**:659-682

[77] Thomsen MG, Galambosi B, Galambosi Z, Uusitalo M, Mordal R, Heinonen A. Harvest time and drying temperature effect on secondary metabolites in *Rhodiola rosea*. Acta Horticulturae. 2012;**955**:243-252

[78] Zhao YH, Jia X, Wang WK, Liu T, Huang SP, Yang MY. Growth under elevated air temperature alters secondary metabolites in *Robinia pseudoacacia* L. seedlings in Cd-and Pb-contaminated soils. Science of the Total Environment. 2016;**565**:586-594

[79] Yadav SK. Cold stress tolerance mechanisms in plants. A review. Agronomy for Sustainable Development. 2010;**30**(3):515-527

[80] Verma N, Shukla S. Impact of various factors responsible for fluctuation in plant secondary metabolites. Journal of Applied Research on Medicinal and Aromatic Plants. 2015;**2**(4):105-113

[81] Janska A, Marsik P, Zelenkova S, Ovesna J. Cold stress and acclimation—What is important for metabolic adjustment? Plant Biology. 2010;**12**:395-405

[82] Chinnusamy V, Zhu J, Zhu J-K. Cold stress regulation of gene expression in plants. Trends in Plant Science. 2007;**12**(10):444-451. DOI: 10.1016/j.tplants.2007.07.002

[83] Esra K, Işlek C, Üstün AS. Effect of cold on protein, proline, phenolic compounds, and chlorophyll content of two pepper (*Capsicum annuum* L.) varieties. Gazi University Journal of Science. 2010;**23**(1):1-6

[84] Griffith M, Yaish MWF. Antifreeze proteins in overwintering plants: A tale of two activities. Trends in Plant Science. 2004;**9**:399-405. DOI: 10.1016/j.tplants.2004.06.007

[85] Bernáth J, Tétényi P. The effect of environmental factors on growth, development, and alkaloid production of poppy (*Papaver somniferum* L.): I. responses to day-length and light intensity. Biochemie und Physiologie der Pflanzen. 1979;**174**:468-478

[86] Janas KM, Cvikrová M, Pałagiewicz A, Szafranska K, Posmyk MM. Constitutively elevated accumulation of phenylpropanoids in soybean roots at low temperatures. Plant Science. 2002;**163**:369-373

[87] Christie PJ, Alfenito MR, Walbot V. Impact of low temperature stress on general phenylpropanoid and anthocyanin pathways: Enhancement of transcript abundance and anthocyanin pigmentation in maize seedlings. Planta. 1994;**194**:541-549

[88] Pedranzani H, Sierra-de-Grado R, Vigliocco A, Miersch O, Abdala G. Cold, and water stresses produce changes in endogenous jasmonates in two populations of Pinus pinaster Ait. Plant Growth Regulation. 2003;**52**:111-116

[89] Lei XY, Zhu RY, Zhang GY, Dai YR. Attenuation of cold-induced apoptosis by exogenous melatonin in carrot suspension cells: The possible involvement of polyamines. Journal of Pineal Research. 2004;**36**:126-131. DOI: 10.1046/j.1600-079X.2003.00106.x

[90] Zhao Y, Qi L, Wei-Ming WW, Saxena PK, Chun-Zhao LC. Melatonin

improves the survival of cryopreserved callus of *Rhodiola crenulata*. Journal of Pineal Research. 2011;**50**:83-88. DOI: 10.1111/j.1600-079X.2010.00817.x

[91] Kovacs Z, Simon-Sarkadi L, Szucs A, Kocsy G. Differential effects of cold, osmotic stress and abscisic acid on polyamine accumulation in wheat. Amino Acids. 2011;**38**:623-631. DOI: 10.1007/s00726-009-0423-8

[92] Nadeau P, Delaney S, Chouinard L. Effects of cold hardening on the regulation of polyamine levels in wheat (*Triticum aestivum* L.) and Alfalfa (*Medicago sativa* L.). Plant Physiology. 1987;**8**:73-77. DOI: 10.1104/pp.84.1.73

[93] Hummel I, EI-Amrani A, Gouesbet G, Hennion F, Couee I. Involvement of polyamines in the interacting effects of low temperature and mineral supply on Pringlea antiscorbutica (Kerguelen cabbage) seedlings. Journal of Experimental Botany. 2004;**399**:1125-1134

[94] Zhong JJ, Yoshida T. Effects of temperature on cell growth and anthocyanin production in suspension cultures of *Perilla frutescens*. Journal of Fermentation and Bioengineering. 1993;**76**(6):530-531

[95] Narayan MS, Thimmaraju R, Bhagyalakshmi N. Interplay of growth regulators during solid-state and liquid-state batch cultivation of anthocyanin producing cell line of *Daucus carota*. Process Biochemistry. 2005;**40**:351-358

[96] Thimmaraju R, Bhagyalakshmi N, Narayan MS, Ravishankar GA. Kinetics of pigment release from hairy root cultures of Beta vulgaris under the influence of pH, sonication, temperature, and oxygen stress. Process Biochemistry. 2003;**38**:1069-1076

[97] Yang L, Wen KS, Ruan X, Zhao YX, Wei F, Wang Q. Response of plant secondary metabolites to environmental factors. Molecules. 2018;**23**(4):E762. DOI: 10.3390/molecules23040762

[98] Mahajan S, Tuteja N. Cold, salinity and drought stresses: An overview. Archives of Biochemistry and Biophysics. 2005;**444**:139-158. DOI: 10.1016/j.abb.2005.10.018

[99] Abideen Z, Qasim M, Rasheed A, et al. Antioxidant activity and polyphenolic content of Phragmites karka under saline conditions. Pakistan Journal of Botany. 2015;**47**(3):813-818

[100] Petrusa LM, Winicov I. Proline status in salt tolerant and salt sensitive alfalfa cell lines and plants in response to NaCl. Plant Physiology and Biochemistry. 1997;**35**:303-310

[101] Parida AK, Das AB. Salt tolerance and salinity effects on plants: A review. Ecotoxicology and Environmental Safety. 2005;**60**:324-349. DOI: 10.1016/j.ecoenv.2004.06.010

[102] Navarro JM, Flores P, Garrido C, Martinez V. Changes in the contents of antioxidant compounds in pepper fruits at ripening stages, as affected by salinity. Food Chemistry. 2006;**96**:66-73

[103] Mutiu F, Bozcuk S. Salinity-induced changes of free and bound polyamine levels in sunflower (*Helianthus annuus* L.) roots differing in salt tolerance. Pakistan Journal of Botany. 2007;**39**:1097-1010

[104] Borgognone D, Cardarelli M, Rea E, Lucini L, Colla G. Salinity source-induced changes in yield, mineral composition, phenolic acids and flavonoids in leaves of artichoke and cardoon grown in floating system. Journal of the Science of Food and Agriculture. 2014;**94**:1231-1237

[105] Ben Abdallah S, Aung B, Amyot L, Lalin I, Lachâal M, Karray-Bouraoui N, et al. Salt stress (NaCl) affects plant growth and branch pathways of carotenoid and flavonoid biosyntheses in *Solanum nigrum*. Acta Physiologiae Plantarum. 2016;**38**(3):72. DOI: 10.1007/s11738-016-2096-8

[106] Salama ZA, Gaafar AA, El-Fouly MM. Genotypic variations in phenolic, flavonoids and their antioxidant activities in maize plants treated with Zn (II) HEDTA grown in salinized media. Agricultural Sciences. 2015;**6**(3):397

[107] Chunthaburee S, Sanitchon J, Pattanagul W, Theerakulpisut P. Effects of salt stress after late booting stage on yield and antioxidant capacity in pigmented rice grains and alleviation of the salt-induced yield reduction by exogenous spermidine. Plant Production Science. 2015;**18**(1):32-42. DOI: 10.1626/pps.18.32

[108] Ashraf MA, Ashraf M, Ali Q. Response of two genetically diverse wheat cultivars to salt stress at different growth stages: Leaf lipid peroxidation and phenolic contents. Pakistan Journal of Botany. 2010;**42**(1):559-565

[109] Ahmed A, Gabr A. AL–Sayed H, Smetanska I, effect of drought and salinity stress on total phenolic, flavonoids and flavonols contents and antioxidant activity in in vitro sprout cultures of garden cress (*Lepidium sativum*). Journal of Applied Sciences Research. 2012;**8**(8):3934-3942

[110] Wahid A, Ghazanfar A. Possible involvement of some secondary metabolites in salt tolerance of sugarcane. Journal of Plant Physiology. 2006;**163**(7):723-730. DOI: 10.1016/j.jplph.2005.07.007

[111] Lim J-H, Park K-J, Kim B-K, Jeong J-W, Kim H-J. Effect of salinity stress on phenolic compounds and carotenoids in buckwheat (*Fagopyrum esculentum* M.) sprouts. Food Chemistry. 2012;**135**(3):1065-1070. DOI: 10.1016/j.foodchem.2012.05.068

[112] Valifard M, Mohsenzadeh S, Kholdebarin B, Rowshan V. Effects of salt stress on volatile compounds, total phenolic content and antioxidant activities of *Salvia mirzayanii*. South African Journal of Botany. 2014;**93**:92-97. DOI: 10.1016/j.sajb.2014.04.002

[113] Xu Z, Zhou G, Shimizu H. Plant responses to drought and rewatering. Plant Signaling & Behavior. 2010;**5**:649-654. DOI: 10.4161/psb.5.6.11398

[114] Ibrahim W, Zhu Y, Chen Y, Qiu CW, Zhu S, Wu F. Genotypic differences in leaf secondary metabolism, plant hormones and yield under alone and combined stress of drought and salinity in cotton genotypes. Physiologia Plantarum. 2019;**165**:343-355

[115] Davis RF, Earl HJ, Timper P. Effect of simultaneous water deficit stress and Meloidogyne incognita infection on cotton yield and fiber quality. Journal of Nematology. 2014;**46**(2):108

[116] Farooq M, Hussain M, Wahid A, Siddique KHM. Drought stress in plants: An overview. In: Plant Responses to Drought Stress: From Morphological to Molecular Features. Berlin, Heidelberg: Springer; 2012. pp. 1-33. DOI: 10.1007/978-3-642-32653-0_1

[117] Ullah A, Manghwar H, Shaban M, Khan AH, Akbar A, Ali U, et al. Phytohormones enhanced drought tolerance in plants: A coping strategy. Environmental Science and Pollution Research. 2018;**25**:33103-33118

[118] Larson RA. The antioxidants of higher plants. Phytochemistry. 1988;**27**:969-978

[119] Anjum F, Yaseen M, Rasul E, Wahid A, Anjum S. Water stress in barley (*Hordeum vulgare L.*). II. Effect on chemical composition and chlorophyll contents. Pakistan Journal of Agricultural Sciences. 2003;**40**:45-49

[120] Massacci ASM, Nabiev L, Pietrosanti SK, Nematov TN, Chernikova K. Thor, Leipner J. Response of the photosynthetic apparatus of cotton (*Gossypium hirsutum*) to the onset of drought stress under field conditions studied by gas-exchange analysis and chlorophyll fluorescence imaging. Plant Physiology and Biochemistry. 2008; 46:189-195

[121] Soliz-Guerrero JB, de Rodriguez DJ, Rodriguez-Garcia R, Angulo-Sanchez JL, Mendez-Padilla G. Quinoasaponins: Concentration and composition analysis. In: Janick J, Whipkey A, editors. Trends in New Crops and New Uses. Alexandria: ASHS Press; 2002. p. 110

[122] Winkel-Shirley B. Flavonoid biosynthesis, a colorful model for genetics, biochemistry, cell biology, and biotechnology. Plant Physiology. 2001;**26**:485-493. DOI: 10.1104/pp.126.2.485

[123] Zhang L, Wang Q, Guo Q, Chang Q, Zhu Z, Liu L, et al. Growth, physiological characteristics and total flavonoid content of *Glechoma longituba* in response to water stress. Journal of Medicinal Plants Research. 2012;**6**(6):1015-1024

[124] Khalid KA. Influence of water stress on growth, essential oil, and chemical composition of herbs (*Ocimum* sp.). International Agrophysics. 2006;**20**(4):289-296

[125] Gharibi S, Tabatabaei BES, Saeidi G, Goli SAH. Effect of drought stress on total phenolic, lipid peroxidation, and antioxidant activity of *Achillea* species. Applied Biochemistry and Biotechnology. 2016;**178**(4):796-809. DOI: 10.1007/s12010-015-1909-3

[126] Moharramnejad S, Sofalian O, Valizadeh M, Asgari A, Shiri M. Proline, glycine betaine, total phenolics and pigment contents in response to osmotic stress in maize seedlings. Journal of Bioscience & Biotechnology. 2015;**4**:313-319

[127] Quan NT, Anh LH, Khang DT, Tuyen PT, Toan NP, Minh TN, et al. Involvement of secondary metabolites in response to drought stress of rice2(*Oryza sativa* L.). Agriculture. 2016;**6**(2):23

[128] Nacer B. Soybean seed phenol, lignin, and isoflavones and sugars composition altered by foliar boron application in soybean under water stress. Food and Nutrition Sciences. 2012;**3**(4):12. DOI: 10.4236/fns.2012.3408

[129] Salem N, Msaada K, Dhifi W, Sriti J, Mejri H, Limam F, et al. Effect of drought on safflower natural dyes and their biological activities. EXCLI Journal. 2014;**13**:1-18

[130] Król A, Amarowicz R, Weidner S. Changes in the composition of phenolic compounds and antioxidant properties of grapevine roots and leaves (*Vitis vinifera* L.) under continuous long-term drought stress. Acta Physiologiae Plantarum. 2014;**36**(6):1491-1499. DOI: 10.1007/s11738-014-1526-8

[131] Ghosh S, Watson A, Gonzalez-Navarro OE, et al. Speed breeding in growth chambers and glasshouses for crop breeding and model plant research. Nature Protocols. 2018;**13**(12):2944-2963. DOI: 10.1038/s41596-018-0072-z

[132] de Castro EM, Pinto JEBP, Bertolucci SKV, Malta MR, Cardoso MG,

Silva FAM. Coumarin contents in young Mikania glomerata plants (guaco) under different radiation levels and photoperiod. Acta Farmaceutica Bonaerense. 2006;**25**(3):387-392

[133] Li TSC, Mazza G, Cottrell AC, Gao L. Ginsenosides in roots and leaves of American ginseng. Journal of Agricultural and Food Chemistry. 1996;**44**(3):717-720. DOI: 10.1021/jf950309f

[134] Binder BYK, Peebles CAM, Shanks JV, San K-Y. Influence of culture medium composition and light conditions on the accumulation of bioactive compounds in shoot cultures of *Scutellaria laterifora* L. (American skullcap) grown in vitro. Applied Biochemistry and Biotechnology. 2017;**183**:1414-1425. DOI: 10.1007/s12010-017-2508-2

[135] Chen CC, Agrawal DC, Lee MR, et al. Influence of LED light spectra on in vitro somatic embryogenesis and LC-MS analysis of chlorogenic acid and rutin in *Peucedanum japonicum* Thunb.: A medicinal herb. Botanical Studies. 2016;**57**(1):1-8. DOI: 10.1186/s40529-016-0124-z

[136] Ali M, Abbasi BH. Light-induced fluctuations in biomass accumulation, secondary metabolites production, and antioxidant activity in the cell suspension cultures of *Artemisia absinthium* L. Journal of Photochemistry and Photobiology B: Biology. 2014;**140**:223-227. DOI: 10.1016/j.jphotobiol.2014.08.008

[137] Liang B, Huang X, Zhang G, Zhang F, Zhou Q. Effect of lanthanum on plants under supplementary ultraviolet-B radiation: Effect of lanthanum on flavonoid contents in soybean seedlings exposed to supplementary ultraviolet-B radiation. Journal of Rare Earths. 2006;**24**:613-616

[138] Shiozaki N, Hattori I, Gojo R, Tezuka T. Activation of growth and nodulation in a symbiotic system between pea plants and leguminous bacteria by near-UV radiation. Journal of Photochemistry and Photobiology B: Biology. 1999;**50**(1):33-37

[139] Ramakrishna A, Dayananda C, Giridhar P, Rajasekaran T, Ravishankar GA. Photoperiod influences endogenous indoleamines in cultured green alga Dunaliella bardawil. Indian Journal of Experimental Biology. 2011;**49**:234-240

[140] Abbasi BH, Tian C-L, Murch SJ, Saxena PK, Liu C-Z. Light-enhanced caffeic acid derivatives biosynthesis in hairy root cultures of Echinacea purpurea. Plant Cell Reports. 2007;**26**:1367-1372

[141] Liu CZ, Guo C, Wang YC, Ouyang F. Effect of light irradiation on hairy root growth and artemisinin biosynthesis of *Artemisia annua* L. Process Biochemistry. 2002;**38**(4):581-585

[142] Regvar M, Bukovnik U, Likar M, Kreft I. UV-B radiation affects flavonoids and fungal colonization in *Fagopyrum esculentum* and *F. tataricum*. Open Life Sciences. 2012;**7**:275-283

[143] Markham KR, Tanner GJ, Caasi-Lit M, Whitecross MI, Nayudu M, Mitchell KA. A possible protective role for 30, 40-dihydroxyflavones induced by enhanced UV-B in a UV-tolerant rice cultivar. Phytochemistry. 1998;**49**:1913-1919

Chapter 4

Types and Function of Phytohormone and Their Role in Stress

Diksha Vaishnav and Parul Chowdhury

Abstract

Plants require sunlight, water, oxygen, and minerals to grow and flourish. Along with the external environments, plant cell functioning is regulated by chemicals and plant hormones, also known as phytohormones or plant growth regulators (PGRs). Plant hormones are chemical substances, like signalling molecules found in plants at extremely low concentrations. Hormones such as auxins, cytokinins, gibberellins, ethylene, abscisic acid, jasmonic acid; salicylic acid, brassinosteroids, and strigolactones are the classes of plant hormones playing vital role in plant. All these hormones are produced in practically every region of the plant and are distributed throughout the plant. Hormones, as well as external variables, play a vital role in processes such as vernalisation, phototropism, seed germination, and dormancy, because these hormones are responsible for translating the external signal into adaptive growth and developmental changes, that help plant to survive better. They also evolved as cellular signal molecules with important roles in the modulation of immunological responses to bacteria, insect herbivores, and beneficial microorganisms. Hence, plant hormones govern a variety of biological activities ranging from growth and development to biotic and abiotic responses. This chapter will focus on various classes of plant hormones and their role in growth and development along with the stress.

Keywords: plant hormone, signalling molecules, growth and development, abiotic and biotic stress

1. Introduction

1.1 The meaning of a plant hormone

Plants need sunlight, water, oxygen, minerals for their growth and development. These are. Apart from external factors there are some intrinsic factors that regulate the growth and development of plants [1]. These are called plant hormones or "Phytohormones". The term "hormone" was originally used narrowly to refer to secretory substances generated by particular organs, glands, tissues, or cells in animals that were transported by veins (or other comparable tissues) to more or less specific tissues and had some effect on their metabolism [2]. However, the terminology is now always used to refer to the mobile signals of living creatures and is frequently used with a heading to indicate the type of living things, such as animal hormones, plant hormones, insect hormones, etc. Every living species has hormones that are frequently exclusive to that species and reflect the traits of its biological occurrences. The peculiar growth phenomenon known as developmental plasticity, in which a plant continuously forms new organs and tissues throughout its life cycle, can be attributed, at least in part, to the features of plant hormones [3]. Plant hormones are a group of organic, naturally occurring chemicals that, when present in small amounts, affect physiological functions. Growth, differentiation, and development are the primary processes impacted, while other processes, such stomatal movement, may also be impacted. Although the term "phytohormones" is not commonly used, plant hormones have also been referred to as such [4].

- Plant hormones are chemical compounds present in very low concentration in plants. They are derivatives of indole (auxins), terpenes (Gibberellins), adenine (Cytokinins), carotenoids (Abscisic acid) and gases (Ethylene). These hormones are produced in almost all parts of the plant and are transmitted to various parts of the plant. They may act synergistically or individually. Roles of different hormones can be complementary or antagonistic. Hormones play an important role in the processes like vernalisation, phototropism, seed germination, dormancy etc. along with extrinsic factors.

- Two types of plant hormone Synthetic and Natural. Synthetic plant hormones are exogenously applied for controlled crop production Charles Darwin first observed the phototropism in the coleoptiles of canary grass and F.W. Went first isolated auxin from the coleoptiles of oat seedlings. Plant hormone (phytohormones) is chemicals produced by plants that regulate their growth, development, reproductive processes, longevity, and even death.

Plant hormones control every phase of the life cycle of the plant. In general, more than one hormone influences plant biological activity, therefore the biological phenomena frequently represents the combined interactions of multiple distinct hormones [5]. When plants confront biotic and abiotic pressures, they can only survive by altering various biological processes, unlike animals that may flee from harsh situations. In these circumstances, plant hormones also work together to alter biological reactions for the establishment and maintenance of plant stress tolerance. Transducing extracellular or intracellular signals into cellular responses is a process known as signal transduction. The processes of signal transduction are involved in hormones functioning. Hormones are compounds that function at low concentrations that circulate through some or all living creatures to signal and regulate the response, growth, and development of those organisms [6].

2. Class of phytohormone

Plant hormones (phytohormones) are chemicals produced by plants that regulate their growth, development, reproductive processes, longevity, and even death. These small molecules are derived from secondary metabolism and are responsible for the adaptation of plants to environmental stimuli. Plants are subjected to an ever changing environment and require these phytohormones for appropriate responses. A single phytohormone can regulate many cellular and developmental processes, while at the same time multiple hormones often influence a single process [7].

Auxin, gibberellins, cytokinins, ethylene and abscisic acid are the five primary phytohormone. Other phytohormone that influences plant physiological processes include brassinosteroids, salicylates, jasmonates, strigolactones, etc (**Table 1**) [8].

2.1 Auxin

2.1.1 Introduction

The first growth hormone to be identified was auxin. They were discovered due to the observations of Charles Darwin and his son, Francis Darwin. The coleoptile (protective sheath) of canary grass develops and bends in the direction of the light source as seen by the Darwin's. This is known as "phototropism." Additionally, their research demonstrated that the coleoptile tip was the location where the bending occurred. As a result, F. W. Went was able to isolate the first auxin from the coleoptile tip of oat seedlings [9]. The apical meristem of shoots, young leaves, and seeds is where auxin is mostly produced. From the point of production, auxin moves downward in a unidirectional or polar manner. Auxin concentration gradient produced by polar transit drives specific responses. The plasma membrane's auxin-specific transport proteins regulate how auxin leaves the call. Plant hormones work through signal transduction, triggering several cellular responses. Auxin attaches to receptors that are related to enzymes, which encourages reaction catalysis. The repressor protein for certain genes (the auxin response gene) attaches to ubiquitin when auxin binds to a receptor. This causes the repressor protein to be degraded, and the transcription of auxin response genes proceeds, promoting cellular growth and development [10].

2.1.2 Function

Auxin, plays a crucial role in regulating growth and development. Indole-3-Acetic Acid (IAA), Indole-3-Butyric Acid (IBA), and 4-chloro-indole-3-acetic acid are all members of this hormone family that are found in nature. Auxin levels vary dramatically within the plant body and throughout the life cycle of the plant, forming complex gradients that appear to be a central component of its regulatory activity for plant development. In order to control auxin levels in particular tissues in response to shifting environmental and developmental factors, plants have evolved complex networks with adaptive flexibility as well as genetic and biochemical redundancy [11]. Indole-3-acetic acid (IAA) is the main auxin in most plants (natural auxin) 2, 4-Dichlorophenoxy Acetic acid, Indole-3-Propionic Acid, alpha- Naphthalene Acetic Acid are synthetic auxins.

2.2 Gibberellins

2.2.1 Introduction

Gibberellins are plant growth regulators that control growth and have an impact on a variety of developmental processes, including stem elongation, germination, blooming, enzyme induction, and so on. The most pronounced effect of gibberellins on plant

Hormone	Where produced or found in plant	Major function
Auxin (IAA)	Shoot apical meristems and young leaves are the primary site of auxin synthesis, root apical meristem also produce auxin, also the root depends on shoot for much of its auxin .developing seeds and fruits contain high levels of auxin, but it is unclear whether it is newly synthesised or transports from maternal tissues	Stimulate stem elongation (low concentration only); promotes the formation of lateral and adventitious roots; regulates development of fruit: enhances apical dominances; functions in phototropism and gravitropism; promotes vascular differentiation; retards leaf abscission.
Cytokinins	These are synthesised primarily in roots and transported to other organs, although there are many minor sites of production as well.	Regulate cell division in shoots and roots; modify apical dominance and promote lateral bud growth; promote movement of nutrients into sink tissues; stimulate seed germination; delay leaf senescence.
Gibberellins	Meristems of apical bud and roots, young leaves, and developing seed are the primary sites of production.	Stimulate stem elongations, pollen development pollen tube growth, fruit growth, and seed development and germination; regulate sex determination and the transition from juvenile to adult phases.
Brassinosteroids	These compounds are present in all plant tissues, although different intermediates predominate in different organs. Internally produced brassinosteroids act near the site of synthesis.	Promote cell expansion and cell division in shoots; promote root growth at high concentrations; inhibit root growth at high concentration; promote xylem differentiation and inhibit phloem differentiation; promote seed germination and pollen tube elongation.
Abscisic acid (ABA)	Almost all plant cells have the ability to synthesise abscisic acid, and its presence has been detected in every major organ and living tissue; may be transported in the phloem or xylem.	Inhibits growth; promotes stomatal closure during drought stress; promotes seed dormancy and inhibits early germination; promotes leaf senescence; promotes desiccation tolerance.
Strigolactones	These carotenoid-derived hormones and extracellular signals are produced in roots in response to low phosphate conditions or high auxin flow from the shoot.	Promote seed germination, control of apical dominance, and the attraction of mycorrhizal fungi to the root.
Ethylene	This gaseous hormone can be produced by most parts of the plant. It is produced in high concentrations during senescence, leaf abscission, and the ripening of some types of fruits. Synthesis is also stimulated by wounding and stress.	Promotes ripening of many types of fruit, leaf abscission, and the triple response in seedling (inhibition of stem elongation, promotion of lateral expansion, and horizontal growth); enhances the rate of senescence; promotes root and root hair formation; promotes flowering in the pineapple family.

Table 1.
Summary of various phytohormone with their major functions.

development is the elongation of the stem [12]. When it is administered to a shrub at low concentration, the stem begins to grow. Different dwarf kinds' genetic restrictions are overcome via Gibberellins. More than 70 gibberellins have been isolated. The numbers are GA1, GA2, GA3, and so forth. The most extensively researched plant growth regulator is GA3, or gibberellic acid. GAs are a family of plant hormones with about 135 members that are classified as diterpenoids with a gibberellin basic structure [13].

2.2.2 Function of gibberellins

- **Seed germination.** In the absence of sunshine, some seeds that are light-sensitive, like lettuce and tobacco, germinate poorly. If the seeds are placed in the sunshine, germination starts right away. The need for light can be met if the seeds are given a gibberellic acid treatment.

- **Dormancy of buds.** Autumn-formed buds dormant till the spring. By administering gibberellin to them, you can break them out of their dormant stage.

- **Root growth.** Gibberellins hardly have any impact on root development. A few plants, nevertheless, may experience modest growth inhibition at a greater dose.

2.3 Cytokinins

2.3.1 Introduction

Adenine derivatives known as cytokinin have the capacity to stimulate cell proliferation in tissue culture [14]. The most common natural occurrence of cytokinin in plants is zeatin. Cytokinin is transported via the xylem from roots to shoots. Exogenous injections of Cytokinins stimulate cell division in tissue culture when auxin is present. Cytokinins promote shoot initiation in moss; Cytokinins induce bud formation Growth of lateral buds. Cytokinin applications, or the increase in Cytokinin levels in transgenic plants with genes for enhanced Cytokinin synthesis, can cause the release of lateral buds from apical dominance [15].

Natural Example of plant hormone Cytokinins are isopentenyl adenine and Zeatin (corn kernels, coconut milk) while Synthetic: Benzyladenine, Kinetin, thidiazuron, and diphenylurea [16].

2.3.2 Functions of plant hormone cytokinins

- This promotes lateral and adventitious shoot growth and is used in culture to initiate shoot production.

- Assists in resolving auxin-induced apical dominance.

- Stimulate the production of chloroplast in the leaves.

- Promoting the mobilisation of nutrients and slowing leaf senescence.

Cytokinin helps in encouraging plant growth and cell division. Utilised by farmers to boost crop output even under drought-like circumstances, it has a positive effect on cotton seedlings by 5–10% seedling emergence [17]. By promoting resistance to certain disease-causing bacteria, plays a significant part in the pathogenesis of plants.

2.3.3 Cytokinin transport

In the xylem and phloem, cytokinins are moved from roots to shoots and the other way around. By conveying information about nutrient availability, for instance, transported cytokinins may play a part in coordinating root and shoot growth. To enable effective mobilisation and tailored translocation of cytokinins, many cellular importers and exporters are necessary, although little is understood about cytokinin transporters. According to transport studies, cytokinins are transported through a widespread H + -coupled high-affinity purine transport pathway [18].

2.4 Ethylene

2.4.1 Introduction

The important hormone ethylene controls and mediates intricate cycles in plants that affect their growth and development as well as their ability to survive throughout their life cycle. The capacity of ethylene to ripen fruits and cause senescence is its primary use and area of scientific study. The potential to accelerate the ripening of fruits where ethylene is the primary hormone, such as tomato and banana fruits, has been the main focus for food biotechnologists. By regulating the manufacture of the ethylene hormone, food biotechnologists hope to be able to control the ripening of fruit [19].

We must first comprehend how ethylene is secreted in the tissues of a plant in order to comprehend its function. Two stages make up the metabolic process that produces ethylene [20].

It begins with a substance called SAM (S-adenosyl-L-methionine). The enzyme ACS aids in the conversion of SAM into ACC (ACC synthase). ACO is an enzyme that converts ACC to ethylene (ACC oxidase) [21].

It is important to realise that the enzymes ACS and ACO are both released by various gene coding families in synchrony with one another when conditions like drought, flood, wound, exerting pressure from the outside, and pathogen assault occur [22].

2.4.2 Functions of ethylene

In plants, ethylene is used for a variety of purposes. Seed germination, shoot and root growth, root development, abscission of leaves and fruits, the creation of adventitious roots, senescence of leaves and flowers, and sex determination are a few of the crucial tasks that ethylene performs. For instance, in plant tissue, ethylene stimulates the development of air-filled cavities known as aerenchym tissues during floods, which aids in the oxygenation of plants. However, the ripening of climacteric fruits, such as peaches, bananas, apples, and tomatoes, is ethylene's most significant role. For instance, putting a ripe banana in a bag of immature avocados would speed up the avocados' ripening process. The build-up of ethylene in the bag is to blame for this.

In summary Ethylene's important uses are the following:

- The generation of female flowers in a male plant.

- Producing root growth to enhance the capability of the root to absorb more water and minerals.

- Evoking a phenomenon called epinasty. Epinasty is a complex behaviour seen in plants when the roots are flooded. During floods, the top layer of the leaves grows more than the bottom ones. This induces the leaves to drop and rather than being horizontal the leaves become more vertical. This is specially induced by ethylene when it is converted to ACC and transported from the xylem to the tissues of leaves on the upper part.

- Ethylene promotes negative geotropism, where it ensures that the growth of the roots is towards the ground. Hence, more area of roots in the soil indicates easy absorption of minerals from the soil.
- The sex of a flower can be determined.
- Influences seed germination.
- Has a great role in the initiation of root growth and pollination.
- The flowering of pineapple flowers can be hastened by ethylene.
- It breaks the dormancy of buds, seeds and storage organs of the plants.
- It increases the dormancy of lateral buds and improves apical dominance [23].

2.5 Abscisic acid

2.5.1 Introduction

Abscisic acid is the plant stress hormone (ABA). It inhibits plant development and regulates abscission and dormancy. The naturally occurring Abscisic acid is dextrorotatory (+), but commercially available synthesised ABA is a racemic combination. ABA is transported by the xylem, phloem, and parenchyma cells [24].

Discovery of ABA took place between 1950 and 1960, scientists had a hunch that when a growth stimulating endogenous hormones are present in the plant cell, growth inhibiting hormones which causes the senescence or abscission of fruits must be governed by other hormones namely abscisic acid(ABA). ABA does not cause abscission, they just inhibit growth [25]. Violaxanthin and neoxanthin are Xanthophylls that are used to synthesise ABA. Epoxidation, or the presence of epoxy-carotenoids, is required for ABA production. The synthesis however initiates from IPP forming GGPP further leads to the formation of Zeaxanthin produces violaxanthin. Violaxanthin forms cis—neoxanthin followed by cis—xanthin produces ABA Aldehyde leads to ABA [26]. Synthesis occurs in mature leaves and stems, as well as developing fruits, seeds.

Abscisic acid is referred to as a stress hormone since its production is promoted by environmental challenges such as drought and water logging. It is crucial in the tolerance of abiotic stress. ABA is important in a variety of developmental and physiological processes, including:

- ABA causes stomata to close when excessive salinity, water stress, and lowers water loss through transpiration. To stimulate stomatal closure, ABA interacts with other phytohormones such as jasmonates, nitric oxide, and signalling molecules.

- ABA causes seed dormancy, allowing seeds to tolerate desiccation and other unfavourable growth factors. Seeds can be stored for an extended period of time.
- ABA is essential for root development and modification under nitrogen deprivation and drought. It controls gene expression, which is necessary for root development, maintenance and water absorption.
- ABA affects protein-encoding genes as well as lipid and storage protein production.
- ABA is required for the signal transduction pathway during the stress response.
- Abscisic acid participates in the production of dehydrins, osmoprotectants, and protective proteins.
- By modulating stress-responsive genes, ABA promotes long-term development [27].

2.6 Jasmonic acid

2.6.1 Introduction

Jasmonic acid (JA) is an organic compound found in several plants including jasmine. The molecule is a member of the jasmonate class of plant hormones. It is biosynthesized from linolenic acid by the octadecanoid pathway. It was first isolated in 1957 as the methyl ester of jasmonic acid by the Swiss chemist Edouard Demole and his colleagues.

Jasmonates are represented by jasmonic acid (JA) and its methyl ester. The plant hormone jasmonic acid (JA) and its derivative (jasmonoyl isoleucine: JA-Ile) are signalling molecules involved in the control of cellular defence and development in plants. Jasmonic acid plays a vital role in the various plant developmental processes including flowering, fruiting, senescence and secondary metabolism. These are known to be critically important in plant defence and abiotic stress response. Jasmonic acid stimulates the antioxidant system, induces amino acid and soluble sugar accumulation, and modulates stomatal opening and closing during abiotic stress [28].

2.6.2 Function

1. Jasminates play an important role in plant defence, where they induce the synthesis of proteinase inhibitors which deter insect feeding, and, in this regard, act as intermediates in the response pathway induced by the peptide system in.
2. Jamonates inhibit many plant processes such as growth and seed germination.
3. They promote senescence, abscission, tuber formation, fruit ripening, pigment formation and tendril coiling.
4. JA is essential for male reproductive development of Arabidopsis. The role in other species remains to be determined [29].

2.7 Salicylic acid (SA)

2.7.1 Introduction

Salicylic acid (SA) is a member of the salicylates group of chemicals, which are phenolic compounds generated by plants and have an aromatic ring and a hydroxyl group. Salicylates were used as pain relievers for thousands of years before they were chemically identified.

Salicylic acid and its derivatives, as one of the plant hormones generated naturally, belong to the phenolic acid group and consist of a ring connected to the hydroxyl and carboxyl groups, with cinnamic acid as the starting component. It is mostly produced in cytoplasmic cells of the plant. Symbolised by the symbol SA called chemical orthohydroxybenzoic acid chemical formula is $C7H6O3$ [30].

2.7.2 Salicylic acid's chemical composition

2.7.2.1 Function

Many physiological and biochemical processes, including photosynthesis, ion absorption, membrane permeability, enzyme activity, flowering, heat generation, and plant growth and development, are influenced by Salicylic acid.

Its effects include, among others, the inhibition of root growth, variation in chlorophyll content, carotenoids, and xanthophylls, increased water use efficiency, improved nitrogen uptake by some species, inhibition of ethylene generation, alteration of plant nutrition, inhibition of the absorption of some substances, and regulation of flowering.

Its effects include, among others, the inhibition of root growth, variation in chlorophyll content, carotenoids, and xanthophylls, increased water use efficiency, improved nitrogen uptake by some species, inhibition of ethylene generation, alteration of plant nutrition, inhibition of the absorption of some substances, and regulation of flowering [31].

2.8 Brassinosteroids

2.8.1 Introduction

Brasinosteroids, the sixth plant hormone after auxin, gibberellins, cytokinin, abscisic acid, and ethylene, are structurally similar to steroid hormones found in animals. Brassinosteroids are important plant hormones that function similarly to animal hormones in a variety of biological processes, such as cell division, cell elongation, root development, photomorphogenesis, stomatal and vascular differentiation, seed germination, immunity, and reproduction. Brassinosteroids are also involved in regulating the metabolism of plant oxidation radicals, ethylene synthesis and root gravitropic response, and have a role in mediating plant responses to stress, such as freezing, drought, salinity, disease, heat and nutrient deficiency. Depending on growth state, this subfamily of hormones controls a wide variety of activities in plant development and responses to environmental challenges. Analogs of these hormones have been demonstrated to significantly boost grain production.

There are at least 70 polyhydroxylated sterols in the class of Brasinosteroids. These substances are similar in structure to animal steroid hormones that control the function of complex genes is influenced by the expression of several genes metabolic processes, which help to control cell division, differentiation, help in morphogenesis regulation, and regulate certain plant development stages, including blooming and cell expansion [32].

2.8.2 Function

Agriculture is where brassinosteroid is initially used. Therefore, during both abiotic and biotic stress, such as salt and drought stress, extremes in temperature, and disease assault, brassinosteroid mediates in plants. By applying them exogenously, they can be utilised to regulate the time of blooming in some plant species. Brassinosteroid deficiency has been linked to dwarfing phenotypes, short petioles, delayed blooming, and decreased fertility in plants [33].

1. Cell Division, possibly by increasing transcription of the gene encoding cyclinD3 which regulates a step in the cell cycle

2. Cell elongation, where BRs promote the transcription of genes encoding xyloglucanases and expansins and promote wall loosening. This leads to stem elongation.

3. Vascular differentiation.

4. BRs are needed for fertility: BR mutants have reduced fertility and delayed senescence probably as a consequence of the delayed fertility.

5. Inhibition of root growth and development.

6. Promotion of ethylene biosynthesis and epinasty [34].

2.9 Strigolactones

2.9.1 Introduction

Strigolactones are a class of carotenoid-derived plant hormones that are found in a wide variety of plant species, ranging from mosses to higher plants. They are crucial for the stimulation and branching of parasitic plants as well as the symbiosis and growth of arbuscular mycorrhizal fungus in soil. In 1966, cotton root exudate was the source of the first Strigolactones to be identified, called Strigol. Its name was connected to the Striga plant genus, a typical parasite of this crop. The creation of adventitious and lateral roots, as well as the induction of secondary growth, acceleration of leaf senescence, promotion of internode growth, and root elongation are all regulated by Strigolactones [35].

2.9.2 Function

Plant roots create a class of chemicals known as strigolactones. Strigolactone encourages the growth of parasitic plants like Strigalutea and other members of the genus Striga that grow in the roots of the host plant. Because they form a mutualistic connection with these plants and offer phosphate and other soil nutrients, strigolactone plays a crucial part in the symbiotic fungi's ability to identify the plant. Strigolactone has a function in the enhancement of lateral root development and root hair elongation, as well as the inhibition of shoot branching in plants [36].

3. Role of phytohormone in stress

Due to their effects on hormonal and nutritional imbalances, stress conditions have a significant negative impact on crop productivity. Some common stresses that have a negative impact on plant growth and development include salinity, drought, heavy metals, nutrient deficiency, and pathogens. These stresses have an impact on plant growth in one way or another. A single stress can negatively impact several plant functions in a variety of ways. For instance, salt reduces plant development by generating hormonal abnormalities, ion toxicity, oxidative stress, nutritional problems, and water stress. Plants adapt specific defence systems to deal with biotic and abiotic challenges in their native soil environments. To each given stimulus, several cellular signalling pathways are triggered [37]. The synthesis of phytohormones is accelerated by these signals. Signalling molecules called phytohormones control physiological and developmental processes in plants. There are several biotic and abiotic factors that influence hormone production, which substantially varies These hormones may have a significant impact on plant growth and development even at very low concentrations. Plant defence against environmental stresses depend heavily on hormonal signalling. The primary function of phytohormone production in plants is resistance to stress. Auxin, cytokinins, ethylene, gibberellins, and abscisic acid are the five main groups of phytohormones. Brassinosteroids, jasmonic acid, salicylic acid, and nitric acid have also been found as chemical messengers present in trace amounts in plants in addition to these well-known plant hormones. These hormones move throughout the plant body via the xylem or phloem transport stream.

The most investigated stress-responsive hormone among them is abscisic acid (ABA), which has a role in a variety of conditions include osmotic, drought, and cold stress. Plant processes including organogenesis, embryogenesis, and the development of vascular tissue are all regulated by auxin. A new class of plant hormones known as brassinosteroids affects processes such as flowering, senescence, and seed germination [38].

4. Abiotic stress and phytohormones

Abiotic stress like salt, low temperature, heat, drought, UV-radiation, elevated CO2, ozone, and heavy metals stress, are the major factors that reduce the growth of plants. There are various adaptations and ways by which plants protect them from the abiotic stress, and one such strategy is phytohormones produced by them. Phytohormones play a major role in some of the important functions in plants in how to respond to external environmental changes [39].

Abscisic acid (ABA), also called as the plant stress hormones as it is the most studied plant hormones under abiotic stress and has the important role in stress tolerance too. ABA activates specific signalling molecules which helps in activating the genes by signal transduction and activates set of genes through abscisic acid-responsive elements (ABRE) and Dehydration Responsive elements (DRE) pathways.in response to ABS LEA and dehydration proteins are also induced [40].

Auxin's primary role is to play an important role in plant growth and development, but auxin also plays an important role in abiotic stress tolerance. Functional genomics studies based upon the transcriptome analysis proves the upregulated genes related to auxins under abiotic stress conditions [41, 42]. Auxin level has been altered and PIN proteins play an important role in all kinds of abiotic stress like salt, drought, heat, heavy metal etc. As auxin plays a role in root architecture, so it helps in abiotic stress response by altering root physiology [43].

Cytokinin has a main role in growth and development like Auxin, but also it helps in stress signalling, which was proved by the transcriptomics studies. At stress forms cytokinin levels increase and decrease accordingly [44]. It was studied that the cytokinin levels initially drop down in less stress conditions but gradually increase as the stress increases. Lots of studies have been done and still more studies need to be done to clarify the statements [45].

Gibberellic acid, which is also called as germinating hormones also plays a role in abiotic stress tolerance. Studies has shown its role in all kinds of stress [46]. Experiments have shown its role in heat stress tolerance in Tomatoes [47], and chilling stress tolerance by increasing antioxidant activity in sprouts has also been proved [48]. Studies in *Arabidopsis thaliana* have shown that GA also helps in mitigating heavy metal stress via modulating antioxidants [49].

Salicylic acid (SA) serves as a key hormone in plant innate immunity, including resistance in both local and systemic tissue upon biotic attacks, hypersensitive responses, and cell death. Salicylic acid plays an important role in the growth and development of the plant for important physiological roles such as increasing the plant's response to stress conditions (biotic and abiotic) by increasing the resistance of the plant to System Acquired Resistance (SAR) by stimulating or changing the internal paper dissection endogenous signalling to withstand a large number of stresses [50]. Salicylic acid acts as a stimulant or transmitter of the cell to withstand environmental stress conditions such as drought, cold, heat, stress of heavy elements, and conditions of ammonia tension and also increases the plant's ability to withstand salt stress salt particularly harmful sodium chloride compound NaCl. This phenolic acid hormone plays important role in the regulation of plant growth, fruit ripening and development. It is involved in pathogenesis-related protein expression. It also has the ability to bind conjugate with some amino acids such as proline and arginine, which increase the plant's effectiveness in resisting environmental stresses and at the same time maintain systemic acquired resistance [51].

Salicylic acid and Jasmonic acid, hormones which are known for the biotic stress hormones also play a role in abiotic stress tolerance. In wheat application of SA and JA has shown to increase the germination during the drought stress [52]. Even exogenous application of SA helps in salt stress response in cucumbers [53] and reports have shown their role in heavy metal toxicity tolerance too [54].

Strigolactones are also very important hormones in plant microbe relations but also play a role in different stress. Moreover Strigolactones play a role in stress adaptation by cross talking with other plant hormones and playing a regulatory role [55].

5. Biotic stress and phytohormones

As soon as plat sense a stress response lots of signalling pathway is triggered in the plant. One of the most important event is alteration in calcium levels in plants, which in turn helps in signalling via PIP and activating the kinases enzymes. Calcium ions bins to calmodulin or calcium-dependent protein kinases (CDPKs), or phosphatases which can either phosphorylate/dephosphorylate specific transcription factors, and regulate the gene expression [56]. Although all plant hormones play a role in stress response both abiotic and biotic as they cross talk among each other and help in combating stress in coordinated manner, but few hormones are well studied and are like font line warriors in biotic stress tolerance, like abscisic acid, ethylene, salicylic acid, jasmonates.

Abscisic acid positively regulates or negatively regulates defence responsive genes depending upon the type of pathogen plant is facing and type of plant. It exerts its effect by upregulating some of the gene families like ABA-responsive elements (ABRE) and other transcription factors like MYC, MYB and NAC families [57]. As the herbivore attacks the plant, their secretions increase the ABA level in the plant. Research has been done in plants with the herbivore attack and mutant analysis too and it proves to be correct. Transcriptome analysis also prove the same [58]. ABA is also involved in plant antiviral defence, as infection with few viruses studied has shown to have increased accumulation of ABA in plants [59].

Salicylic acid, Jasmonic acid and Ethylene play an important role in biotic stress response. These are the key hormones which play an important role in plant defence against the pathogens and pests. Increased Salicylic acid levels protect the whole plants with the help of upregulation of Pathogenesis related genes (PR genes) [60]. Jasmonic acid helps in production of secondary metabolites like tannins, total phenols, total flavonoids, and lignin upon insect attack and helps in insect resistance in plants [61].

All these hormones and signalling pathways interact among each other to help plants fight against the stress. Plant adaptations and its development stages and cross talk of these biotic and abiotic stresses with help of different transcription factors help a plant to survive in harsh conditions.

6. Conclusion

A well-known fact is that plant growth, development and senescence are under the regulation of the system of natural growth regulators: natural inhibitors and phytohormones. These substances could be transported in the cell or even between the plant tissues and organs. Phytohormones are involved in wide range of functions, ranging from growth and development to biotic and abiotic stress tolerance. Efficiency of plant hormone is increased by their ability to crosstalk under different conditions. We can use research in plant hormones to improve crop growth and production, as they regulate complex signalling networks involving developmental processes and environmental stresses. Significant progress has been made in identifying the key components and understanding plant hormone signalling (especially salicylic acid, Jasmonic acid and Ethylene) Several recent studies provide evidence for the involvement of other hormones such as Abscisic acid, Auxin, Gibberellins, Cytokinins and Brassinosteroids in plant defence signalling pathways. To understand how plants coordinate multiple hormonal components in response to various developmental and environmental cues is a major challenge in research. It is important to note that the type of interactions and plant responses to stresses vary depending on the plant system as well as the time, quantity and the tissue where hormones are produced.

Author details

Diksha Vaishnav and Parul Chowdhury*
Dr. B. Lal Institute of Biotechnology, Jaipur, Rajasthan

*Address all correspondence to: parul@drblal.com

References

[1] Lynch J, Marschner P, Rengel Z. Effect of internal and external factors on root growth and development. 3rd ed. In: Marschner's Mineral Nutrition of Higher Plants. 2012. pp. 331-346

[2] Park J, Lee Y, Martinoia E, Geisler M. Plant hormone transporters: What we know and what we would like to know. BMC Biology. 2017;**15**(1):93

[3] Weyers JDB, Paterson NW. Plant hormones and the control of physiological processes. New Phytologist. 2001;**152**(3):375-407

[4] Acharya BR, Assmann SM. Hormone interactions in stomatal function. Plant Molecular Biology. 2008;**69**(4):451-462

[5] Wang YH, Irving HR. Developing a model of plant hormone interactions. Plant Signaling & Behavior. 2011;**6**(4):494-500

[6] Totaro A, Panciera T, Piccolo S. YAP/TAZ upstream signals and downstream responses. Nature Cell Biology. 2018;**20**(8):888-899

[7] Peres A, Soares J, Tavares R, Righetto G, Zullo M, Mandava N, et al. Brassinosteroids, the sixth class of phytohormones: A molecular view from the discovery to hormonal interactions in plant development and stress adaptation. International Journal of Molecular Sciences. 2019;**20**(2):331

[8] Gray WM. Hormonal regulation of plant growth and development. PLoS Biology. 2004;**2**(9):e311

[9] Hajný J, Prát T, Rydza N, Rodriguez L, Tan S, Verstraeten I, et al. Receptor kinase module targets PIN-dependent auxin transport during canalization. Science. 2020;**370**(6516):550-557

[10] Chandler JW. Auxin response factors. Plant, Cell & Environment. 2016;**39**(5):1014-1028

[11] Lavy M, Estelle M. Mechanisms of auxin signaling. Development. 2016;**143**(18):3226-3229

[12] Torii KU, Hagihara S, Uchida N, Takahashi K. Harnessing synthetic chemistry to probe and hijack auxin signaling. New Phytologist. 2018;**220**(2):417-424

[13] Kawaguchi M, Syono K. The excessive production of Indole-3-acetic acid and its significance in studies of the biosynthesis of this regulator of plant growth and development. Plant and Cell Physiology. 1996;**37**(8):1043-1048

[14] Bonner J, Bandurski RS. Studies of the physiology, pharmacology, and biochemistry of the auxins. Annual Review of Plant Physiology. 1952;**3**:59-86

[15] Brian PW. Effects of gibberellins on plant growth and development. Biological Reviews. 1959;**34**(1):37-77

[16] Yamaguchi I, Cohen JD, Culler AH, Quint M, Slovin JP, Nakajima M, et al. 4.02–plant hormones. In: Liu HW, Mander L, editors. Comprehensive Natural Products II, Chemistry and Biology. 2010. pp. 9-125

[17] Camara MC, Vandenberghe LPS, Rodrigues C, de Oliveira J, Faulds C, Bertrand E, et al. Current advances in gibberellic acid (GA3) production, patented technologies and potential applications. Planta. 2018;**248**(5):1049-1062

[18] Kieber JJ, Schaller GE. Cytokinins. The Arabidopsis Book/American Society of Plant Biologists. 2014;**12**(2014)

[19] Davies PJ. The plant hormones: Their nature, occurrence, and functions. In: Plant Hormones. Dordrecht: Springer; 2010. pp. 1-15

[20] Khan AA. Cytokinins: Permissive role in seed germination: With other plant hormones, cytokinins regulate germination and dormancy by a novel mechanism. Science. 1971;**171**(3974):853-859

[21] Shanthi V. Actinomycetes: Implications and Prospects in Sustainable Agriculture. Biofertilizers: Study and Impact; 2021. pp. 335-370

[22] Kieber JJ, Schaller GE. Cytokinin signaling in plant development. Development. 2018;**145**(4):dev149344

[23] El-Showk S, Ruonala R, Helariutta Y. Crossing paths: Cytokinin signalling and crosstalk. Development. 2013;**140**(7):1373-1383

[24] Schaller GE, Kieber JJ. Ethylene. The Arabidopsis book/American Society of Plant Biologists. 2002;**1**:1

[25] Abeles FB, Morgan PW, Saltveit ME Jr. Ethylene in Plant Biology. Academic Press; 2012

[26] Sauter M, Moffatt B, Saechao MC, Hell R, Wirtz M. Methionine salvage and S-adenosylmethionine: Essential links between sulfur, ethylene and polyamine biosynthesis. Biochemical Journal. 2013;**451**(2):145-154

[27] Tiwari S, Gupta D, Fatima A, Singh S, Prasad SM. Phytohormones and their metabolic engineering for abiotic stress. pp. 541-574

[28] Crawford BL Jr, Lancaster JE, Inskeep RG. The potential function of ethylene. The Journal of Chemical Physics. 1953;**21**(4):678-686

[29] Moirangthem K, Tucker G. How do fruits ripen. Frontiers for Young Minds. 2018;**6**:16

[30] Bhatla SC. Jasmonic Acid. Development and Metabolism: Plant Physiology; 2018

[31] Mueller MJ. Enzymes involved in jasmonic acid biosynthesis. Physiologia Plantarum. 1997;**100**(3):653-663

[32] Raskin I. Role of salicylic acid in plants. Annual Review of Plant Biology. 1992;**43**(1):439-463

[33] Chen Z, Zheng Z, Huang J, Lai Z, Fan B. Biosynthesis of salicylic acid in plants. Plant Signaling & Behavior. 2009;**4**(6):493-496

[34] Hassoon AS, Abduljabbar IA. Review on the Role of Salicylic Acid in Plants. Sustainable Crop Production. IntechOpen; 2019. pp. 61-66

[35] Filgueiras CC, Martins AD, Pereira RV, Willett DS. The ecology of salicylic acid signaling: Primary, secondary and tertiary effects with applications in agriculture. International Journal of Molecular Sciences. 2019;**20**(23):5851

[36] Clouse SD. Brassinosteroids. The Arabidopsis Book/American Society of Plant Biologists. 2011;**9**:9

[37] Bajguz A, Tretyn A. The chemical characteristic and distribution of brassinosteroids in plants. Phytochemistry. 2003;**62**(7):1027-1046

[38] Ali B. Practical applications of brassinosteroids in horticulture—Some field perspectives. Scientia Horticulturae. 2017;**225**:15-21

[39] Bajguz A, Hayat S. Effects of brassinosteroids on the plant responses to

environmental stresses. Plant Physiology and Biochemistry. 2009;**47**(1):1-8

[40] Ruyter-Spira C, Al-Babili S, Van Der Krol S, Bouwmeester H. The biology of strigolactones. Trends in Plant Science. 2013;**18**(2):72-83

[41] Brewer PB, Koltai H, Beveridge CA. Diverse roles of strigolactones in plant development. Molecular Plant. 2013;**6**(1):18-28

[42] Yoneyama K, Xie X, Yoneyama K, Takeuchi Y. Strigolactones: Structures and biological activities. Pest Management Science: Formerly Pesticide Science. 2009;**65**(5):467-470

[43] Fahad S, Hussain S, Bano A, Saud S, Hassan S, Shan D, et al. Potential role of phytohormones and plant growth-promoting rhizobacteria in abiotic stresses: Consequences for changing environment. Environmental Science and Pollution Research. 2015;**22**(7):4907-4921

[44] Nadeem SM, Ahmad M, Zahir ZA, Kharal MA. Role of phytohormones in stress tolerance of plants. In: Plant, Soil and Microbes. Cham: Springer; 2016. pp. 385-421

[45] Sreenivasulu N, Harshavardhan VT, Govind G, Seiler C, Kohli A. Contrapuntal role of ABA: Does it mediate stress tolerance or plant growth retardation under long-term drought stress? Gene. 2012;**506**(2):265-273

[46] Hu W, Zuo J, Hou X, Yan Y, Wei Y, Liu J, et al. The auxin response factor gene family in banana: Genome-wide identification and expression analyses during development, ripening, and abiotic stress. Frontiers in Plant Science. 2015;**6**:742

[47] Van Ha C, Le DT, Nishiyama RIE, Watanabe YASUKO, Sulieman S, Tran UT, et al. The auxin response factor transcription factor family in soybean: Genome-wide identification and expression analyses during development and water stress. DNA Research. 2013;**20**(5):511-524

[48] Kazan K. Auxin and the integration of environmental signals into plant root development. Annals of Botany. 2013;**112**(9):1655-1665

[49] Nishiyama R, Watanabe Y, Fujita Y, Le DT, Kojima M, Werner T, et al. Analysis of cytokinin mutants and regulation of cytokinin metabolic genes reveals important regulatory roles of cytokinins in drought, salt and abscisic acid responses, and abscisic acid biosynthesis. The Plant Cell. 2011;**23**(6):2169-2183

[50] Zwack PJ, Rashotte AM. Interactions between cytokinin signalling and abiotic stress responses. Journal of Experimental Botany. 2015;**66**(16):4863-4871

[51] Banerjee A, Roychoudhury A. The regulatory signaling of gibberellin metabolism and its crosstalk with phytohormones in response to plant abiotic stresses. In: Plant Signaling Molecules. Woodhead Publishing; 2019. pp. 333-339

[52] Guo T, Gull S, Ali MM, Yousef AF, Ercisli S, Kalaji HM, et al. Heat stress mitigation in tomato (*Solanum lycopersicum* L.) through foliar application of gibberellic acid. Scientific Reports. 2022;**12**(1):1-13

[53] Hu Z, Weijian L, Yali F, Huiquan L. Gibberellic acid enhances postharvest toon sprout tolerance to chilling stress by increasing the antioxidant capacity during the short-term cold storage. Scientia Horticulturae. 2018;**237**:184-191

[54] Zhu XF, Jiang T, Wang ZW, Lei GJ, Shi YZ, Li GX, et al. Gibberellic

acid alleviates cadmium toxicity by reducing nitric oxide accumulation and expression of IRT1 in Arabidopsis thaliana. Journal of Hazardous Materials. 2012;**239**:302-307

[55] Ilyas N, Gull R, Mazhar R, Saeed M, Kanwal S, Shabir S, et al. Influence of salicylic acid and jasmonic acid on wheat under drought stress. Communications in Soil Science and Plant Analysis. 2017;**48**(22):2715-2723

[56] Radhakrishnan R, Lee IJ. Regulation of salicylic acid, jasmonic acid and fatty acids in cucumber (*Cucumis sativus* L.) by spermidine promotes plant growth against salt stress. Actaphysiologiaeplantarum. 2013;**35**(12):3315-3322

[57] Aftab T, Khan M, Idrees M, Naeem M, Hashmi N. Methyl jasmonate counteracts boron toxicity by preventing oxidative stress and regulating antioxidant enzyme activities and artemisinin biosynthesis in *Artemisia annua* L. Protoplasma. 2011;**248**(3):601-612

[58] Cheng X, Ruyter-Spira C, Bouwmeester H. The interaction between strigolactones and other plant hormones in the regulation of plant development. Frontiers in Plant Science. 2013;**4**:199

[59] Aldon D, Mbengue M, Mazars C, Galaud JP. Calcium signalling in plant biotic interactions. International Journal of Molecular Sciences. 2018;**19**(3):665

[60] Cao FY, Yoshioka K, Desveaux D. The roles of ABA in plant–pathogen interactions. Journal of Plant Research. 2011;**124**(4):489-499

[61] Tooker JF, De Moraes CM. Feeding by a gall-inducing caterpillar species alters levels of indole-3-acetic and abscisic acid in *Solidago altissima* (Asteraceae) stems. Arthropod-Plant Interactions. 2011;**5**(2):115-124

Chapter 5

Perspective Chapter: Effect of Low-Temperature Stress on Plant Performance and Adaptation to Temperature Change

Veena Devi, Amanpreet Kaur, Mehak Sethi and Gosangi Avinash

Abstract

Low-temperatures (LT) stress is one of the abiotic stresses in plants that affect cell survival, cell division, photosynthesis, and water transport, negatively affecting plant growth, and eventually constraining crop productivity. LT stress is categorized as, (i) chilling stress where low temperature (0–15°C) causes injury without ice crystal formation in plant tissues, and (ii) freezing stress (<0°C), where ice formation occurs within plant tissues. Both stresses are together termed low temperature or cold stress. In general, plants originating from tropical and subtropical regions are sensitive to LT, whereas temperate plants showed chilling tolerance to variable degrees. Low-temperature stress negatively impacts plants, may affect the survival rate of crop plants, and also affect various processes, including cell division, photosynthesis, plant growth, development, metabolism, and finally reduce the yield of crop plants, especially in the tropics and subtropics. To overcome stress generated by low-temperature exposure, plants trigger a cascade of events that enhance their tolerance by gene expression changes and activation of the ROS scavenging system, thus inducing biochemical and physiological modifications. In this chapter, a detailed discussion of different changes in plants and their tolerance mechanism is done to understand the plant's response under LT stress.

Keywords: low-temperature stress, oxidative stress, resilience, stress tolerance

1. Introduction

Low-temperature (LT) stress is one of the abiotic stresses [1] in plants that affect cell survival, cell division, photosynthesis, and water transport with a negative effect on plant growth, eventually constraining crop productivity [2, 3]. LT stress is categorized as, (i) chilling stress, where low temperature (0–15°C) causes injury without ice crystal formation in plant tissues, and (ii) freezing stress (<0°C), where ice formation occurs within plant tissues. Both stresses are termed low temperature or cold stress [4]. In general, plants originating from tropical and subtropical regions are sensitive to LT, whereas temperate plants showed chilling tolerance to variable degrees [2]. Low temperature negatively impacts plants, may affect the survival rate of crop plants, and also affect various processes including cell division, photosynthesis, plant growth, development, metabolism, and finally reduce the yield of crop plants, especially in the tropics and subtropics [5, 6]. To overcome stress generated by LT exposure, plants trigger a cascade of events that enhance their tolerance by changes in gene expression and activation of the ROS scavenging system and thus inducing biochemical and physiological modifications [7, 8]. This review is a detailed discussion of different changes in plants and their tolerance mechanism in order to understand the plant's response under LT stress.

2. Morpho-physiological changes in crop plants in response to LT stress

Morphological changes are the change that is visible on the plants during the early stage of LT stress. These are the primary signs of the plants, indicating adverse effects of stress on plants. Stress reduces leaf expansion, causes chlorosis, wilting of leaves and necrosis, and accelerates senescence in crop plants [9, 10]. Various metabolic reactions were inhibited by LT exposure, consequently preventing the plant's full genetic expression potential expressed by diverse phenotypic symptoms [11]. Low temperature is a limiting factor for seed germination and plant growth [12–14]. Under LT stress in *Elymus nutans* Griseb, the shoot and root lengths in tolerant seedlings were longer than the susceptible ones. Low temperatures also increased the mortality percentage of seedlings [15]. LT dramatically affects photosynthesis as well [16]. The negative impact of abiotic stress on the photosynthetic process in plants has been extensively studied and measurement of chlorophyll fluorescence (Fv/Fm) has proven as an effective, reproducible, and nondestructive tool for evaluating the susceptibility index of plants subjected to LT stress [3, 17]. Under LT stress, photosynthesis is impaired, resulting in a lower amount of carbohydrates for grain production and reducing growth, adding to indirect yield loss [3]. In rice seedlings, LT stress affected total chlorophyll (Chl) content and thus photosynthetic efficiency [18, 19]. Proteomic analysis in a semihardy winter wheat cultivar under natural field conditions indicated a down-regulation of several photosynthesis-related proteins (such as oxygen-evolving enhancer protein, NADH dehydrogenase, and dehydroascorbate reductase) during the initiation of cold acclimation [16]. Low temperature decreases photosynthesis due to partial stomatal closure, slowdown of electron transport, inhibits metabolism of carbohydrates, and interferes with phloem loading [13]. In plants, the content of both total Chl and chlorophyll b (Chl b) decreased and the Chl a/b ratio increased under low night temperature stress [17]. Low night temperature probably enhanced the activity of chlorophyllase enzyme in leaves and hence resulted in reduced Chl synthesis. Most of the Chl a, all the Chl b, and other pigments absorb light. They transfer that light energy to the reaction center but only a part of Chl a molecule can utilize that energy to perform the charge separation process. Plants maintain a relatively higher level of Chl a content, so that they can perform the process of photosynthesis normally and adapt themselves to cold stress. The cessation of growth ensuing from cold stress decreases the capacity of utilizing the energy and hence results in feedback inhibition of photosynthesis. In cold-acclimated winter annuals, Calvin cycle enzymes accumulate in higher amounts and effectively maintain the photosynthetic activity of plants. The Chl content and photosynthetic parameters like Fv/Fm had a positive correlation with chilling injury indices and have been utilized as a marker of cold tolerance in sugarcane [20]. Under dark chilling treatment, Fv/Fm significantly decreased in plants and after the recovery period, the Fv/Fm ratio recovered to nearly that of the control levels [21, 22]. A greater decrease in Chl content in the cold-sensitive rice genotype was reported as compared to the cold-tolerant genotype under cold stress [23]. LT tolerant lines of rice, after stress, seedling height in both the lines remained unchanged over time; however, more tolerant seedlings (M202) exhibited a small increase in the root-to-shoot ratio [23].

Carotenoids are not considered photosynthetic pigments, but play important role in protecting the photosystems from damage. They have structural roles and act as natural antioxidants, quenching triplet Chl and singlet oxygen species, which are potentially harmful to the chloroplast [24, 25]. Carotenoids also maintain and stabilize thylakoid membranes from the damage caused by lipid peroxidation and cold stress [26]. In *Elymus nutans* seedlings, carotenoid content was decreased when exposed to cold stress at 5°C. The decrease in carotenoid content was higher in GN (more sensitive) than DX (tolerant) seedlings [15].

In conclusion, under LT stress plants showed various phenotypic symptoms, these are the primary symptoms of stress. Photosynthetic pigments and photosynthetic parameters like Fv/Fm ratio are altered under LT stress and showed a positive correlation with the chilling injury indices and potential to be used as a marker for cold resistance.

3. Oxidative stress

Plants exposed to LT stress undergo various metabolic and physiological changes and chilling stress ultimately leads to oxidative stress in plants, a physiological condition, where an imbalance occurs between the generation of reactive oxygen species [27] and their metabolism *via* enzymatic and nonenzymatic antioxidants [28]. Different types of reactive oxygen species (ROS) are accumulated under LT stress, which includes (a) singlet oxygen ($^1O_2$), (b) superoxide radical ($O_2^{\bullet-}$), (c) hydrogen peroxide (H_2O_2), and (d) hydroxyl radical ($OH^{\bullet}$) [29]. In plant cells, ROS are continuously produced as a

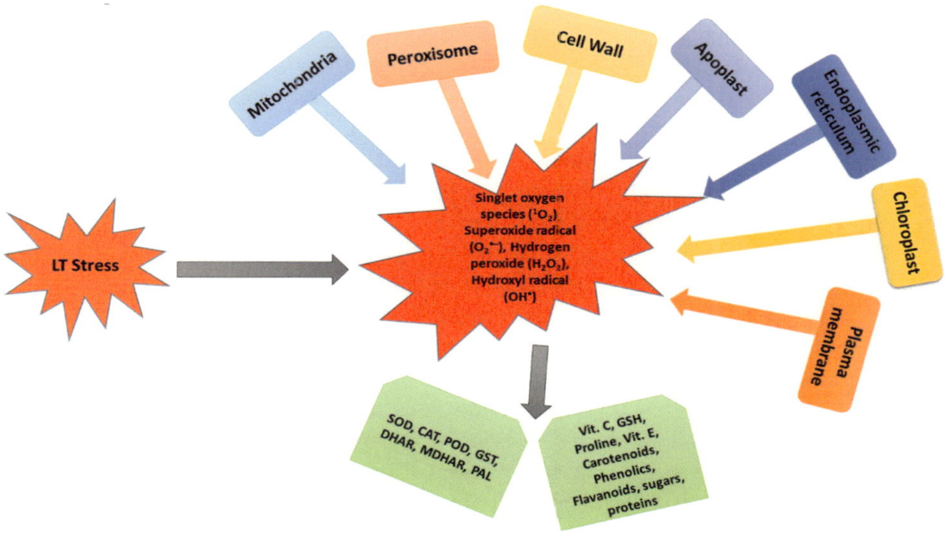

Figure 1.
Overview of oxidative stress, production of reactive oxygen species, and its scavenging.

consequence of aerobic metabolism in all the intracellular organelles, particularly in the chloroplast, mitochondria, and peroxisomes [30]. Chloroplast is considered the main source of ROS in plants. Other ROS-producing sources include NADPH oxidases, cell wall-bound peroxidases, and amine oxidases (**Figure 1**).

4. Other biochemical changes

Under normal physiological conditions, ROS levels are maintained low by the action of various enzymatic and nonenzymatic antioxidants, such as superoxide dismutase (SOD), catalase (CAT), peroxidase (POD), glutathione (GSH), and vitamin C [31]. Accumulation of ROS accelerated under extremely cold conditions, beyond the plant's tolerant level due to less activity of antioxidant enzymes, which are responsible for detoxification of ROS. Higher content of ROS causes oxidative stress which is manifested as peroxidation of membrane lipids, damage to proteins, carbohydrates, and DNA, etc. [28, 32, 33]. They also alter enzyme activities, biochemical reactions, and plant processes, such as photosynthesis and respiration, which negatively affect the plant's survival percentage [12].

ROS alters the activities of enzymes and affects various biochemical reactions and physiological processes, including nutrient movements, respiration, photosynthesis, and transpiration, thus having a negative impact on a plant's survival percentage. Higher H_2O_2 accumulation in cold-stressed leaves of chickpea plants resulted in membrane injury [34]. Oktem *et al.* [35] also stated that an increase in oxidative damage caused by cold stress in lentils resulted due to high H_2O_2 production. Higher MDA content and higher electrolyte leakage from cell membranes of sensitive plants indicate injury caused by LT stress [36, 37]. Increased content of ROS and malondialdehyde (MDA) under LT stress probably impair metabolism in rice seedlings [38]. A significant increase in lipid peroxidation, membrane leakage, and hydrogen peroxide levels was observed in wheat seedlings subjected to chilling stress [39]. Apostolova *et al.* [40] reported a 40 and 100% increase in the content of H_2O_2 in the leaves of winter wheat and spring wheat, respectively under cold stress. Janmohammadi *et al.* [41] reported that during cold stress less cold-hardy spring wheat cultivar had a higher accumulation of hydrogen peroxide than the winter wheat cultivar. LT stress resulted in increased electrolyte leakage in the leaves of *Avena nuda* L. (naked oats) seedlings. Electrolyte leakage also increased with the prolongation of the stress period [42]. Membranes are a primary site of cold-induced injury because of their central role in the regulation of various cellular processes [43, 44]. LT stress leads to the destruction of cell membrane structure in maize plants [45], change the permeability of membranes, and causes leakage of cell electrolytes [5] and thus damages the plants. It has been demonstrated that LT responses are triggered by membrane rigidification, coupled with calcium influxes, cytoskeletal rearrangements, and the activation of MAPK cascades [46]. ROS are not only the toxic by-products of metabolism but also act as signaling molecules that transform the expression of different genes, for example, genes encoding for antioxidant enzymes and modulators of H_2O_2 production. ROS plays a vital role in plant stress acclimation [47, 48].

In conclusion, ROS are accumulated under LT, which alter the activities of enzymes, affect various biochemical, and physiological processes, and thus affect the plant's survival. Enzymatic and nonenzymatic antioxidants enhance their content under LT stress and are involved in the detoxification of ROS, thus increasing the resistance against the stress condition.

5. Enzymatic antioxidants

Plants have developed ROS scavenging mechanisms, which include a variety of nonenzymatic and enzymatic defense systems to protect cellular membranes and organelles from the damaging effects of ROS [49, 50]. Types of antioxidants produced in the plants are represented in **Table 1**. The degree of damage by ROS depends on the balance between the accumulation of ROS products and their detoxification by the antioxidant scavenging system [49].

The efficiency of the antioxidant defense system to scavenge ROS largely decides the plant's sensitivity to chilling [27, 28, 54]. A higher amount of H_2O_2 produced during stress is detoxified by APX, POD, and CAT in different organelles [48]. Catalase converts H_2O_2 into O_2 and water. Zhao *et al.* [54] reported in tomato cultivars that higher activities of CAT, APX, POX, and SOD could be positively correlated with chilling tolerance. The CAT activity increased in plants under prolonged LT stress [55]. Fahimirad *et al.* [56] recorded an increased CAT activity in canola cultivars in response to LT stress. The increase in activity was higher in winter canola than LT-sensitive spring canola. The LT stress resulted in enhanced peroxidase activity in naked oats (*Avena nuda* L.) [42]. Dai *et al.* [57] observed that after 72 hours of the recovery period, in barley seedlings, the peroxidase activity was significantly higher in the cold-tolerant cultivar ($M_0$103) in comparison to the cold-sensitive cultivar (Chumai). Aydin *et al.* [58] reported that in tomato plants (*Lycopersicum esculentum* L.) highest MDA production occurred after 10 days of stress and SOD enzyme activity gradually increased with increasing exposure to cold stress. Expression of the *SOD* gene and enzyme plays a key role to provide resistance in tomato plants against cold stress. Zhang *et al.* [59] observed that in *C. Sativus,* activities of antioxidant enzymes *viz.* SOD, POD, CAT, and APX were reduced after chilling exposure. Fahimirad *et al.* [56] reported that cold stress exposure enhanced SOD activity by 2.5-fold in winter canola (tolerant) leaves when compared to controls, whereas spring canola (LT sensitive) cultivar showed a 1.7-fold increase. Sun *et al.* [60] reported that in sugar cane seedling roots at 4°C, SOD activity was higher in cold tolerant (GT28) variety than cold-sensitive (ROC22) variety. Various studies showed a similar response to cold stress in wheat [61], strawberries [62], and barley [63]. Hajiboland and Habibi [64] reported that in cold-treated seedlings, the activity of SOD increased significantly, while in the acclimated seedlings, SOD activity did not differ from the control.

CAT and POD are important enzymes that scavenge H_2O_2 [65]. Generally, there is a positive correlation between stress tolerance and the activity of POD, CAT, and SOD enzymes in plants [65]. Javadian *et al.* [61] reported that cold-tolerant wheat cultivars had higher CAT activity. Fahimirad *et al.* [56] reported that winter canola had a greater increase in CAT activity than LT-sensitive spring canola under LT stress.

Antioxidant	Types
Enzymatic antioxidants	Catalase (CAT), ascorbate peroxidase (APX), superoxide dismutase (SOD), glutathione peroxidase (GPX), glutathione reductase (GR), glutathione *S*-transferase (GST), dehydroascorbate reductase (DHAR), monodehydroascorbate reductase (MDHAR), etc. [13, 51–53].
Nonenzymatic antioxidant	Ascorbic acid (Vitamin C), α-tocopherol (Vitamin E), glutathione (GSH), carotenoids, phenolics, and flavonoids, etc. [13, 51–53].

Table 1.
List of the different types of enzymatic and nonenzymatic antioxidants of plant.

Morsy *et al.* [66] reported that under cold stress no change was recorded in peroxidase activity in cold-tolerant as well as cold-sensitive rice seedlings. Liu *et al.* [42] reported that POD activities in naked oats (*Avena nuda* L.) were higher under LT than normal temperature. But with time POD activities decreased greatly, indicating that LT had affected POD enzyme synthesis. Dai *et al.* [57] reported that in two contrasting cold-tolerant cultivars of barley, the tolerant cultivar (M0103) had significantly higher peroxidase activity than the sensitive cultivar (Chumai) after 72 h recovery in cold-treated plants. POD activity increased in *Cucumis sativus*, tomato, and canola under LT stress [61, 67, 68]. Sun *et al.* [60] reported that under LT stress at 4°C POD activity was increased in the roots of sugarcane seedlings than in control. The increase in POD activity was higher in the cold-tolerant genotype (GT28) than cold-sensitive genotype (ROC22). Higher POD and SOD activity probably suggest their possible role in mitigating adverse environmental damage. Hajiboland and Habibi [64] reported a slight increase in CAT activity in both wheat cultivars under chilling temperatures with and without acclimation. In contrast, POD activity increased in spring wheat cultivar but not in winter wheat by both temperature treatments. Gong *et al.* [69] reported that in maize seedlings, cold acclimation enhanced the activity of five antioxidant enzymes catalase (CAT), superoxide dismutase (SOD), ascorbate peroxidase (APX), guaiacol peroxidase (GPX), and glutathione reductase (GR). In conclusion, enzymatic antioxidants accumulate under LT stress and are actively involved in the detoxification of ROS thus enhancing the resistance of the plants.

6. Nonenzymatic antioxidants

In plants ascorbic acid (AsA) and glutathione (GSH) are low molecular weights, nonenzymatic antioxidants, abundantly present, and participate in ROS scavenging [28, 70]. The tripeptide glutathione (γ-glutamyl-cysteinyl-glycine) is widely distributed in plant cells and is implicated in the adaptation of plants to environmental stresses, such as extreme temperatures [48]. It is an important antioxidant associated with the regeneration of AsA in the ascorbate-glutathione cycle and participates in the removal of H_2O_2 [67]. Its antioxidant activity is mainly due to its redox buffer property. It functions to remove toxic peroxides formed in the cell during normal and stressed conditions [70, 71].

$$2GSH + ROOH \rightarrow GSSH + H_2O + R - OH \tag{1}$$

Glutathione detoxifies ROS in concert with NADPH. At low nonfreezing temperatures, several plants accumulate GSH and show an increase in GR activity, indicating a possible role in enhancing chilling tolerance and cold acclimation. A differential elevation in GSH has been reported in a number of LT-exposed plants, including cucumber genotypes [72, 73].

Ascorbic acid (AsA) is one of the universal nonenzymatic water-soluble antioxidants having a substantial potential of scavenging ROS in plants both under stressed and non-stressed conditions [74]. Cell cytoplasm constitutes the most abundant pool of ascorbate, while to some extent it is also transported across the plasma membrane (usually 5%) to the apoplast [75, 76]. Ascorbate is a component of the NADPH/glutathione/ascorbate cycle that removes photosynthetically generated O^{2-} and H_2O_2. It may also directly reduce O^2–, quench 'O_2 and regenerate reduced tocopherol. Lukatkin and Anjum [77] reported that AsA and GSH have a high potential for sustainably increasing chilling resistance in plants. A significant increase in the levels of these antioxidants as well as the activity of NADPH-generating dehydrogenases have been caused by LT stress [78]. The AsA content was more in tolerant chickpea (*Cicer arietinum*) genotypes after chilling at the reproductive phase [79]. Kim *et al.* [80] reported that changes in GSH content in two rice cultivars were not evident until 10 days of cold stress. Ascorbic acid content increased significantly in stressed IR50 seedlings in comparison to the control while M-202 stressed seedlings showed little or no change. Overexpression of SlGMEs (*Solnaum lycopersicon,* GDP-Mannose 3', 5′-epimerase) was reported to cause AsA accumulation with enhanced cold tolerance in tomatoes [81]. Airaki *et al.* [78] reported that in pepper plants LT stress caused a significant increase in the level of soluble nonenzymatic antioxidants; ascorbate and glutathione. Kader *et al.* [39] reported an increase in GSH and free ascorbate content in 15 days old seedlings of two wheat varieties after cold treatment. Esra *et al.* [82] reported proline accumulation in two pepper variety seedling leaves under cold stress as compared to control counterparts. Yadegari *et al.* [83] reported that proline content increased more under acclimation than non-acclimated seedlings of soybean and hence provide more tolerance. Zuther *et al.* [84] reported that proline content was higher in acclimated leaves of *Arabdopsis thaliana* than in non-acclimated leaves and recovered back to normal levels after de-acclimation. Airaki *et al.* [78] reported that in pepper plants LT stress significantly increased the levels of soluble nonenzymatic antioxidants; ascorbate and glutathione. Kim *et al.* [80] reported that LT stress at 9°C resulted in increased proline and glutathione content in IR50 rice seedlings, compared to controls, and change in glutathione content was evident on the 10th day of LT stress. Kim *et al.* [23] also reported similar changes, under LT stress for proline, glutathione, and ascorbic acid in rice seedlings. Zhang *et al.* [85] reported that to resist the effect of cold stress, resistant sugarcane varieties showed a higher accumulation of proline content in leaves than sensitive varieties. Sun *et al.* [60] reported an accumulation of proline content in sugarcane seedlings under cold stress. Krol *et al.* [86] reported that cold stress caused a decrease in the radical scavenging activity in the leaves of both varieties of grapes and the more-tolerant variety was characterized by better scavenging activity. In conclusion, nonenzymatic antioxidant accumulates in plants under LT stress and are involved in the detoxification of ROS, thus enhancing the resistance of plant against stress.

Phenylalanine ammonia-lyase (PAL) is the key enzyme of the phenylpropanoid pathway, converting L-phenylalanine (substrate) into *trans*-cinnamic acid, a precursor of phenolics. The activity of the PAL enzyme increases in response to LT stress [87] and is considered to be one of the main lines of cell acclimation in plants against stress [88]. Phenolics protect plants against ROS by acting as antioxidants [89, 90]. Christopoulos and Tsantili [91] used a PAL inhibitor to prove the role of PAL in the accumulation of phenolics under cold stress. Chilling stimulates the expression of genes for phenylalanine ammonia-lyase (PAL) in cucumber seedlings [92]. Olenichenko *et al.* [93] studied the effect of cold stress on phenolic compounds in winter wheat (*Triticum aestivum* L.) leaves, which resulted in hardening and detected an increased level of phenolic compounds. In chilling stressed petunia leaves, it was observed that stress led to elevated antioxidant capacity and total phenolic content [94]. Hajiboland and Habibi [64] reported that PAL activity was increased in winter wheat cultivars under acclimation and more phenolic content accumulated in seedling leaves. The transcription level of PAL and phenolic content was higher in acclimated chickpea seedlings than in non-acclimated ones [37]. Chilling stimulates the enzymatic activities and the expression of genes for phenylalanine ammonia-lyase (PAL) in cucumber seedlings [92]. Pennycooke *et al.* [94] reported that chilling stress leads to elevated total phenolic content and antioxidant capacity in petunia. Krol *et al.* [86] reported that cold stress caused a decrease in the phenolic content in the leaves of two varieties of grapes, though the more-tolerant variety was characterized by higher phenolic contents. Cold acclimation resulted in a higher accumulation of phenolics, which were positively correlated with the antioxidant capacity of plants. Flavonoids are a type of phenolics, that accumulated at higher rates in leaves and stems of LT-stressed plants, which are responsible for enhanced cold tolerance [95]. Ahmed *et al.* [96] reported that anthocyanin (a flavonoid) content increased in *Brassica rapa* under cold stress. Total phenols are also the components of the nonenzymatic antioxidant system and their content has been correlated with the stress tolerance capacity of plants [97]. Esra *et al.* [82] reported that in pepper (*Capsicum annum* L.) phenolics accumulated in seedling leaves under LT stress. In acclimated plants, enhanced PAL activity and accumulation of different phenolics are thought to play an important role in creating cold tolerance [37]. Schulz *et al.* [98] showed that flavonoid accumulation increased in *Arabidopsis thaliana* after cold acclimation and all acclimated plants performed better under cold stress.

Plants accumulate a variety of compatible solutes, including sugars, polyamines, glycine betaine, and proline, in response to cold and other osmotic stresses [13]. In LT-tolerant plants, such as barley, rye, winter wheat, grapevine, potato, chickpea, and *A. Thaliana*, a positive correlation between improved cold tolerance and accumulation of endogenous proline content was observed [99–101]. In order to enhance the stress tolerance level of plants, proline act as a mediator of osmotic adjustment, proteins, and membrane stabilizer, an osmotic stress-related genes inducer, and a ROS scavenger, so that plants can perform better under stress [99, 100, 102]. The most feasible roles of proline are to (a) maintain the acidity of cytosol, (b) maintain the $NAD^+/NADH$ ratio, (c) enhance photosynthetic efficiency of the photosystem II, and (d) inhibit peroxidation of membrane lipids [103, 104]. Proline accumulated in chilling stressed soybean seedlings [83]. Kim *et al.* [80] reported that in two rice cultivars (IR50 and M-202) proline content increased significantly in stressed IR50 seedlings in comparison to control seedlings, whereas in M-202, stressed seedlings showed little or no change. Cold-acclimated plants recovered faster than non-acclimated plants because of the higher accumulation of proline in acclimated plants.

Fernandez *et al.* [105] reported that carbohydrate metabolism has greater LT sensitivity than other photosynthetic components. Although the precise function of soluble sugars remains unclear, their accumulation in plants under a cold acclimation process suggests that sugars probably play an important role as signaling molecules, cryoprotectants, or osmoregulator [106]. Ruelland *et al.* [107] reported that sugars possess a positive correlation with cold stress tolerance. Sugars under LT stress contribute to preventing the water within the plant cells to freeze because of its typical compatible osmolyte property, hence reducing the availability of water for the ice nucleation process in the apoplast. Sugars replaced water molecules in establishing hydrogen bonds with lipid molecules and hence protecting plant cell membranes during cold-induced dehydration [107]. In addition to these, sugars may also play a role in scavenging reactive oxygen species and contribute to enhanced stabilization of membranes [108, 109]. Hormone signaling and sugar signaling are closely associated processes, which contribute to managing plant growth, development, and defensive responses against stress [110]. Seedling resistance against cold was enhanced when rice seedlings were pretreated with fructose or glucose [66]. Trehalose possesses a unique capacity for reversible water absorption and appears to be superior to other sugars in protecting biological molecules from desiccation-induced damage [111]. Transgenic *A. thaliana* plants showed enhanced freezing tolerance due to the accumulation of trehalose during cold treatment [112].

Sucrose accumulated in higher amounts in LT-stressed plants [113]. Sucrose acts as an osmoprotectant, as it maintains the turgor pressure of cells and stabilizes cell membranes by interacting with phosphate in their lipid headgroups, decreasing membrane permeability [106]. In some plants, the increase in sucrose content can be as high as 10-fold. Lower amounts of other free sugars like glucose and fructose also get accumulated under stress. The LT exposure also leads to fructan synthesis in temperate grasses, which were reported to depend upon sucrose accumulation. The effects were studied on a less cold-hardy spring cultivar (pishtaz) and a cold-hardy winter cultivar (CDC-ospray) of wheat under cold acclimation (20 days at 4°C), interrupted by de-acclimation (10 days at 25°C) and then followed by re-acclimation conditions (10 days at 4°C). Hardening conditions induced the accumulation of carbohydrates in both cultivars and the de-acclimated plants exhibited a significant reduction [114]. Total soluble sugars, reducing sugars, and sucrose contents were higher in cold-acclimated than those in non-acclimated plants of sweet cherry [115].

Accumulation of carbohydrates under LT may be due to enhanced expression and post-translation activation of enzymes of the sucrose synthesis pathway [116] and fructose-1,6-bisphosphatase [115]. Sun *et al.* [60] also reported that in sugarcane seedling roots, soluble sugar content increased after LT stress and increased to a higher value in a cold-tolerant variety of sugarcane. Hajiboland and Habibi [64] reported that under cold stress and acclimation total soluble sugar content increased equally in winter wheat while sugar content was higher in acclimated spring wheat seedlings than in non-acclimated seedlings. Parteli *et al.* [117] reported in 1-year-old plants of coffee under a cold acclimation period, the soluble sugars accumulated and enhanced cold tolerance. Burchett *et al.* [118] reported that in cold-acclimated (at 5°C) winter barley plants, the sugar concentration was slightly lower than in non-acclimated plants. Sugars had a positive correlation with cold stress tolerance because they act as osmolytes and protect the water within the plant cells and reduce water accessibility for ice formation. Sugars also establish hydrogen bonds with lipids by replacing water molecules and hence protect the membranes during cold-induced dehydration. Sugars also act as ROS scavengers and play role in membrane stabilization [107]. Sucrose synthase (Sus) is one of the key enzymes involved in sucrose synthesis metabolism, especially in non-photosynthetic tissues. The reversible transformation of sucrose and UDP into UDP-glucose and fructose is catalyzed by the sucrose synthase enzyme. Under normal growth conditions, Sus activity has been linked to phloem loading-unloading and nodule function [115]. The differential regulation of stress-responsive *Sus* genes in leaves might represent part of a general cellular response to the allocation of carbohydrates during acclimation processes, such as the synthesis of cell walls and starch. Under normal physiological conditions, sucrose synthase has a very low level of expression and serves no apparent metabolic function. In leaves and various organs of plants, stress resulted in the stimulation of the expression of *Sus* gene(s) and enhanced stress tolerance. Turhan and Ergin [115] studied the effect of cold acclimation in sweet cherry. The activity of sucrose synthase was higher in the non-acclimated stage than those in the cold-acclimated stage. Klotz and Haagenson [119] studied the effect of cold stress on sugar beet roots and reported that sucrose synthase enzyme activity showed several-fold changes. Abdel-Latif [120] reported that cold shock in wheat seedlings caused an increase in sucrose synthase enzyme activity. The accumulation of sucrose in cane sugar exposed to LT stress supports the role of this sugar as an osmoprotectant that stabilizes cellular membranes and maintains turgor pressure [121]. Yue *et al.* [122] reported that after cold acclimation total soluble sugars and specific sugars, including glucose, sucrose, and fructose, were constantly elevated during cold acclimation and decreased after de-acclimation in tea plants. Cowie *et al.* [123] reported that in *Arabdopsis thaliana* sucrose had a regulatory role in the acclimation of whole plants to cold and this may be important during diurnal dark periods. Zuther *et al.* [84] reported that sucrose content was higher in acclimated leaves of *Arabdopsis thaliana* than in non-acclimated leaves and recovered back to non-acclimated levels after de-acclimation. Burchett *et al.* [118] also observed that in winter barley plants acclimated at 5/−1°C; day/night, and there was a significant increase in the glucose, sucrose, and fructose content. The sucrose content increased by 4-fold in comparison to non-acclimated plants.

Low-temperature stress resulted in the synthesis of different types of proteins [13]. Proteins are the major players in most cellular events and are directly involved in plant LT responses [124]. Cold stress increased soluble protein content in pepper (*Capsicum annum* L.) varieties [82]. Different plant species have shown that cryoprotective proteins are encoded by a range of cold-induced genes. Specific proteins with antifreeze activity (antifreeze proteins, AFPs), accumulated during cold acclimation in the apoplast, thus enhancing plant resistance against freezing stress [125–127]. These AFPs were identified as chitinase-like proteins, β-1,3-glucanase-like proteins, thaumatin-like proteins, and polygalacturonase inhibitor proteins [127, 128]. They were also present in non-acclimated plants, but at different locations and did not exhibit antifreeze activity, which suggested that different isoforms of PR proteins are produced under LT. Until now, no plant has been reported to have constitutive antifreeze activity. However, different studies reported the accumulation of transcripts and translation products of AFP genes during cold acclimation [128]. A number of studies have shown that after exposure to LT, many PR genes get induced and enhanced disease resistance was observed in plants [129].

Xu *et al.* [130] found that frost-sensitive winter wheat cultivars exhibited high levels of ROS and leaf cell death in response to abrupt freezing stress, whereas significant increases in the relative abundance of antioxidant-related proteins were found in frost-tolerant cultivar leaves. Under LT stress in sugarcane seedling roots, the total soluble protein was higher in the cold-tolerant variety than cold-sensitive variety [60] and helped to tolerate LT stress. Moieni-Korbekandi [52] reported in canola (*Brassica napus* L.) seedling leaves that soluble protein content increased under cold stress. Esra *et al.* [82] reported that in two pepper (*Capsicum annum* L.) varieties total soluble protein content was higher under cold stress conditions. These proteomic results emphasize the assumption that freezing-tolerant plants are capable of managing ROS-mediated damage more efficiently than sensitive ones. Sarhadi *et al.* [131] investigated the interrelationship between vernalization fulfillment and expression of LT-induced proteins in wheat genotypes differing in freeze tolerance. Their results showed a clear induction of cold-regulated (Cor)/Lea and antifreeze proteins (AFPs) during cold acclimation in the freeze-tolerant genotype, whereas less induction was observed in the semi-hardy genotype. In winter rye seedlings one of the cold-induced thermal hysteresis proteins was β-1,3-glucanase [132]. Consequently, these proteins must possess extensive structural similarities with the pathogen-induced basic β-1,3-glucanase in tobacco. Cryoprotection increased linearly with an increase in β-1,3-glucanase concentration. Chang *et al.* [133] reported that the protein in the cell sap of cold-acclimated mungbean seedlings was 60% higher than control seedlings.

Yadegari *et al.* [83] reported that total protein content increased in both acclimated and non-acclimated seedlings of soybean.

Proteins with antifreeze activity were reported to be accumulated in the apoplast during cold acclimation, thereby offering plant resistance against freezing [127]. Winter rye antifreeze proteins (AFPs) enhance freezing tolerance by preventing physical damage caused by ice crystals and may also function as a barrier to inhibit ice formation [126]. These proteins were identified as β-1,3-glucanase-like proteins, and chitinase-like proteins [127, 128]. Their results interestingly revealed that during the cold acclimation process, the production of ice nucleation substances in both the leaf and the crown was suppressed, correlating with the rapid up-regulation of genes encoding the major antifreeze (chitinases, glucanases, and thaumatin-like proteins) and ice recrystallization inhibition proteins. Antifreeze proteins (AFPs) or ice recrystallization inhibition (IRI) proteins ascribe to a category of proteins in plants that allow their survival in sub-zero situations. Sarhadi *et al.* [131] showed the expression of LT-induced proteins in wheat genotypes differing in freeze tolerance. Their results clearly showed the induction of cold-regulated (Cor)/Lea and antifreeze proteins (AFPs) during cold acclimation in the freezing-tolerant genotype. Comparable results were also observed in diploid wild wheat (*Triticum urartu* L.), where cold acclimation increased the abundance of ROS-scavenging proteins, LEA/RAB proteins, and dehydrins [134].

Species adapted by natural selection to LT environments have evolved a number of morphological, physiological, and biochemical means to improve survival under prolonged LT stress periods [135]. Cold-adapted species generally have short stature, small leaf surface area, and a high root /shoot ratio. Seedlings subjected to prolonged LT exposure showed chlorosis, wilting, reduced leaf expansion, necrosis, tissue

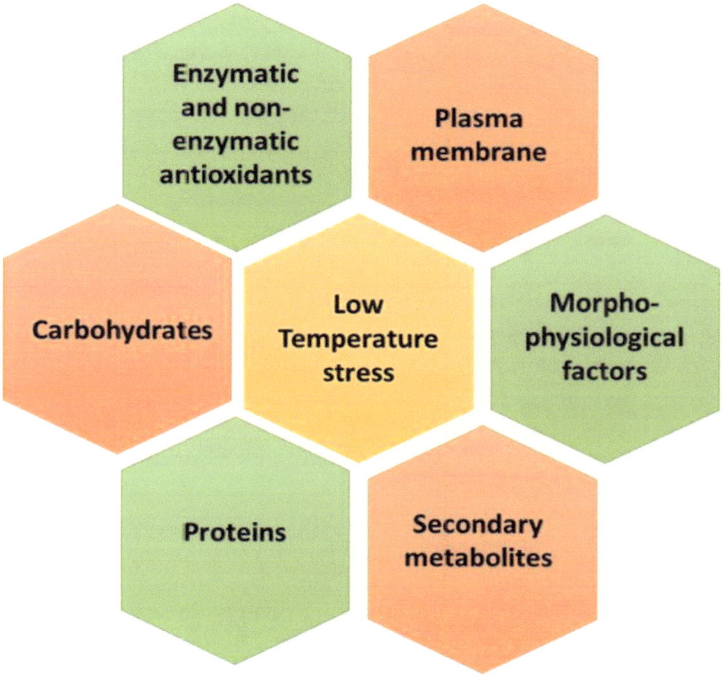

Figure 2.
Factors affected under low-temperature stress.

damage, and stunting [23]. Numerous studies indicated that an increase in antioxidants positively correlated with tolerance to LT stress in plants [23, 28, 136]. Fahimirad *et al.* [56] reported that winter canola had higher activity of antioxidant enzymes (SOD, CAT, and APX) and lower levels of MDA as compared to spring canola. The study showed a positive correlation between the activities of antioxidant enzymes and cold tolerance in the canola winter cultivar as compared to the spring cultivar. Sato *et al.* [137] reported that under cold stress, rice plants protected themselves from oxidative damage by increased production of various antioxidant species. Dai *et al.* [57] reported that LT-treated barley cultivars showed an increase in peroxidase activity after 72 hours of the recovery period. The activity of peroxidase in the tolerant cultivar ($M_0$103) was significantly higher than in the sensitive cultivar (Chumai). Liu *et al.* [42] reported that LT tolerance in *Avena nuda* L. was probably due to the higher content of proline, SOD, CAT, and POD activities. Cold stress conditions, caused a slight decrease in Fv/Fm ratio in plants that showed tolerance to cold, but a significant decrease was observed in plants that are sensitive to LT [138, 139]. Zhang *et al.* [140] reported that under 0°C treatment, plantlets of the tolerant genotype of strawberry showed a significant increase in peroxidase activity (**Figure 2**).

7. Cold acclimation

In several species, the acquisition of freezing tolerance can be induced by exposure to low, nonfreezing, and non-injurious temperatures [124, 141]. Acclimation may be defined as changes that occur in a plant in response to chilling temperatures, which confer subsequent tolerance to the cold injury [113], especially during germination and early seedling growth [69]. Cold priming/acclimation is associated with multiple physiological and biochemical alterations, including membrane stabilization, increased ROS and methylglyoxal (MG) detoxifications, activation of cold-sensitive protein kinases, NO and hormone biosynthesis, and accumulation of antioxidants, HSPs, cold-regulated proteins (CORs), and dehydrins [141–149]. Cold acclimation makes plants capable of protecting themselves from freezing-induced injury [149, 150]. Gong *et al.* [69] reported that in maize seedlings, cold acclimation resulted in higher survival percentage, catalase, ascorbate peroxidase, superoxide dismutase activity, and lower electrolyte leakage, than in non-acclimated seedlings. Cold acclimation in *Zoysia spp.* resulted in higher ABA and H_2O_2 levels as well as regulated antioxidant metabolism, resulting in improved freezing tolerance [151]. Cold priming-induced proline and glycine betaine accumulations were found to be associated with freezing tolerance in barley and *Arundo donax* L. [152]. Cold acclimation in *Arabidopsis thaliana* L. induced the accumulation of endogenous NO, and increased proline levels, conferring freezing tolerance [54]. Cold acclimation also enhanced the expression of genes that play role in membrane stabilization against freeze-induced damage [153, 154]. Minami *et al.* [155] verified that plasma membrane subfractions, responded to cold, by considerably changing lipid and protein composition in Arabidopsis plants. The study indicated that the plasma membrane is restructured in order to resist different stresses that take place throughout a freeze–thaw cycle. Cold acclimation increased the abundance of ROS-scavenging proteins, LEA/RAB proteins, and dehydrins in diploid wild wheat (*Triticum urartu* L.) [134, 156]. Studies have suggested that the activity of cold/chilling-induced genes may facilitate the metabolic changes that confer LT tolerance [156, 157]. Cold acclimation causes the synthesis of protective molecules, such as soluble sugars, sugar alcohols, proline, and glycine betaine [106]. These molecules in conjunction with various proteins play a role to stabilize both phospholipids and proteins of the membranes and proteins of cytoplasm, maintain hydrophobic interactions between molecules and scavenge various types of ROS, which are produced under LT [158]. Some plants respond to LT by the synthesis of some specific proteins that are similar to plant pathogen-related (PR) proteins (particularly in winter rye), in response to cold and drought [159].

Kim *et al.* [23] reported that 14 days of LT stress killed most of the IR50 (sensitive) rice seedlings, while no negative effect was observed in M202 (tolerant) seedlings. Morsy *et al.* [66] showed that cold-tolerant seedlings of rice had 100% survival at 13/10°C regime in comparison to cold-sensitive seedlings, which suffered 50% mortality under the same conditions. Gong *et al.* [69] also reported that the percentage survival of maize seedlings increased after the pretreatment of seedlings at 1°C. Kargiotidou *et al.* [160] reported that the percentage survival of cotton is enhanced if plants are acclimated at low and nonfreezing temperatures prior to cold stress. Many studies in the literature have reported that LT stress affected seedling growth parameters viz. germination, and seedling growth, and caused chlorosis, wilting of leaves, reduced leaf expansion, and necrosis of tissue [12–14]. Low-temperature stress in *Elymus nutans* Griseb decreased shoot lengths of tolerant (DX) and sensitive (GN) genotypes by 88.8 and 91.7%, respectively compared to controls [15]. Jan *et al.* [161] showed that under cold stress one variety of rice SB showed no change in shoot length, while B-385 showed a slight decrease in average shoot length. Both the varieties showed some increase in root length under cold stress. Razmi *et al.* [162] reported that in sorghum (*Sorghum bicolor* L.) genotypes, LT reduced the germination percentage, root length, and shoot length of the seedlings, whereas increased the root/shoot ratio. Increased RL/SL under LT might be an indication of water deficit stress due to cold stress.

LT stress increased the chlorophyllase enzyme activity in grapevine leaves and restrained the synthesis of total Chl [163]. Plants need to maintain a sufficient level of Chl a content to perform the photosynthesis process to some extent even under stress [17]. Yadegari *et al.* [83] reported that under LT stress at 5°C in soybean seedlings Chl a, b, and total Chl contents decreased. However, Yang *et al.* [17] reported that in bitter gourd genotypes, Chl a content increased, whereas total Chl and Chl b contents decreased under cold stress. Our results are supported by the work of Esra *et al.* [82] who reported that in two pepper varieties (*Capsicum annum* L.), Mert and KM-121, the content of Chl a and total Chl significantly decreased, while no significant change was found in the content of Chl b in response to LT stress. Tang *et al.* [20] also reported that under low LT stress in different sugarcane genotypes total Chl content decreased. Carotenoids are not considered photosynthetic pigments, but they play an important role in the protection of the photosystems and accumulate under LT. Carotenoids act as natural antioxidants by quenching triplet Chl and singlet oxygen species, which are potentially harmful to the chloroplast [24, 25]. Fu *et al.* [15] reported that under LT stress at 5°C in *Elymus nutans* seedlings, carotenoid content decreased. The decrease in carotenoid content was higher in GN (more sensitive) than DX (tolerant) seedlings. Gerganova *et al.* [164] reported that in tomato plants after cold treatment, a pronounced decrease was observed in carotenoids. Yadegari *et al.* [83] reported that Chl a, b, and total Chl decreased in both acclimated and non-acclimated soybean seedlings, but in cold-acclimated leaves, this decrease was lesser than in non-acclimated seedlings. It is well documented that photosynthetic apparatus is sensitive to several environmental stresses and PS II appears to be preferentially affected by chilling stress [165]. Fv/Fm reflects the susceptibility to damage of the photosystem II (PSII). Yang *et al.* [17] reported that in two bitter gourd genotype seedling leaves, the Fv/Fm ratio was hardly affected by cold stress, suggesting that LT did not affect the efficiency of PSII. Grapevine seedlings grown under LT stress showed different effects on photosynthetic efficiency [166]. Tang *et al.* [20], however, reported that LT stress in different sugarcane genotypes affected Fv/Fm ratio, which decreased with temperature and stress period. Decreased Fv/Fm ratio indicated that the photosystem was affected under LT stress in sugarcanes. Many researchers pointed out that there was a significant decrease in Fv/Fm ratio under dark chilling stress and after the recovery period the values recovered to control levels [21, 22]. Mishra *et al.* [167] reported that Fv/Fm decreased in both acclimated and non-acclimated samples of *Arabadopsis thaliana*, but the decrease was more in sensitive than tolerant types. Hajiboland and Habibi [64] reported that cold and acclimation both did not affect the Fv/Fm of winter wheat "Sabalan" while causing a significant reduction of *Fv/Fm* in "Zagros" spring wheat. Khaledian *et al.* [37] reported an accumulation of H_2O_2 under cold stress in the leaves of chickpea plants.

Yang *et al.* [17] also reported higher electrolyte leakage for the sensitive bitter gourd genotype (Y-106-5) than the less sensitive one (Z-1-4). Liu *et al.* [42] reported an increase in electrolyte leakage in leaves of *Avena nuda* L. (naked oats) seedlings with cold stress and with prolongation of the stress period. LT stress leads to the destruction of cell membrane structure in maize plants [45], which caused increased permeability of membranes, and increased leakage of cell electrolytes and thus causing damage to plants. Electrolyte leakage was significantly (CD at 5%) lower in acclimated seedlings under LT stress, which probably suggested that membranes of acclimated seedlings were less affected under LT stress. Gong *et al.* [69] reported that electrolyte leakage from root tips of non-acclimated maize seedlings significantly increased after exposure to chilling stress, while cold shock pre-treatment remarkably reduced the leakage of electrolytes under chilling stress as compared to non-acclimated. Aaron *et al.* [168] reported that cold acclimation enhanced the freezing tolerance in *Petunia hybrida* and decreased the EL_{50} value. In conclusion, acclimation prior to LT stress results in the enhanced tolerance of plants. Different types of molecules accumulate under stress conditions, which are used as a potential acclimatizing agent for plants in the form of a spray.

8. Conclusion

The LT stress negatively impacts the plant's performance, and survival percentage, through the generation of ROS. ROS accumulate under the stress in plants from different cell organelles. To deplete these ROS plants, activate the defense system, which includes enzymatic and nonenzymatic antioxidants. These together are involved in the detoxification of the accumulated ROS and enhance resilience against cold stress. Carbohydrates, proteins, and phenolic compounds also accumulate under stress conditions, helping in scavenging the ROS species. Acclimation is a method in which the plants are allowed to be exposed under nonfreezing and non-injurious temperatures, which leads to certain changes in plants that confer subsequent tolerance to cold injury.

Author details

Veena Devi[1*], Amanpreet Kaur[2], Mehak Sethi[1] and Gosangi Avinash[2]

1 ICAR-Indian Institute of Maize Research, Ludhiana, Punjab, India

2 Department of Biochemistry, Punjab Agricultural University, Ludhiana, Punjab, India

*Address all correspondence to: veenyadav.rj@gmail.com

References

[1] Shi Y, Tian S, Hou L, Huang X, Zhang X, Guo H, et al. Ethylene signalling negatively regulates freezing tolerance by repressing expression of CBF and type-A ARR genes in Arabidopsis. The Plant Cell. 2012;**24**:2578-2595

[2] Sanghera GS, Wani SH, Hussain W, Singh NB. Engineering cold stress tolerance in crop plants. Current Genomics. 2011;**12**:30-43

[3] Jena KK, Kim SM, Suh JP, Yang CI, Kim YJ. Identification of cold-tolerant breeding lines by quantitative trait loci associated with cold tolerance in rice. Crop Science. 2012;**51**:517-523

[4] Singh BK, Sutradhar M, Singh AK, Mandal N. Cold stress in rice at early growth stage—An overview. International Journal of Pure Applied Bioscience. 2017;**5**:407-419

[5] Lukatkin AS, Brazaityte A, Bobinas C, Duchovskis P. Chilling injury in chilling-sensitive plants: A review. Zemdirbyste-Agriculture. 2012;**99**:111-124

[6] Li ZG, Zeng HZ, Ao PX, Gong M. Lipid response to short-term chilling shock and long-term chill hardening in Jatropha curcas L. seedlings. Acta Physiologiae Plantarum. 2014;**36**:2803-2814

[7] Hasanuzzaman M, Hossain MA, Teixeira D, Silva JA, Fujita M. Plant responses and tolerance to abiotic oxidative stress: Antioxidant defenses is a key factors. In: Bandi V, Shanker AK, Shanker C, Mandapaka M, editors. Crop Stress and its Management: Perspectives and Strategies. Berlin: Springer; 2013. pp. 261-316

[8] Megha S, Basu U, Kav NNV. Regulation of low temperature stress in plants by microRNAs. Plant, Cell & Environment. 2017;**2017**:1-2

[9] Nahar K, Biswas JK, Shamsuzzaman AMM, Hasanuzzaman M, Barman HN. Screening of indica rice (*Oryza sativa* L.) genotypes against low temperature stress. Botanical Research International. 2009;**2**:295-303

[10] Soufi S, Rezgui S, Bettaeib T. Early effects of chilling stress on the morphological and physiological statut of pretreated Stevia rebaudiana Bert. Seedlings Journal of New Science Agricultural Biotechnology. 2015;**14**:467-472

[11] Chinnusamy V, Zhu J, Zhu JK. Cold stress regulation of gene expression in plants. Trends in Plant Science. 2007;**12**:444-451

[12] Solanke AU, Sharma AK. Signal transduction during cold stress in plants. Physiology and Molecular Biology of Plants. 2008;**14**:69-79

[13] Ruelland E, Zachowski A. How plants sense temperature. Environmental and Experimental Botany. 2010;**69**: 225-232

[14] Nahar K, Biswas JK, Shamsuzzaman AMM. Cold Stress Tolerance in Rice Plant: Screening of Genotypes Based on Morphophysiological Traits. Berlin: LAP Lambert Academic; 2012

[15] Fu JJ, Liu J, Yang LY, Miao YJ, Xu YF. Effects of low temperature on seed germination, early seedling growth and antioxidant systems of wild Elymus nutans grised. Journal of Agricultural Science and Technology. 2017;**19**:1-8

[16] Janmohammadi M, Matros A, Mock HP. Proteomic analysis of cold

acclimation in winter wheat under field condition. Icelandic Agricultural Sciences. 2014;**27**:3-15

[17] Yang J, Kong Q, Xiang C. Effects of low night temperature on pigments, Chl a fluorescence and energy allocation in two bitter gourd (*Momordica charantia* L.) genotypes. Acta Physiologiae Plantarum. 2009;**31**:285-293

[18] Kanneganti V, Gupta AK. Overexpression of OsiSAP8, a member of stress associated protein (SAP) gene family of rice confers tolerance to salt, drought and cold stress in transgenic tobacco and rice. Plant Molecular Biology. 2008;**66**:445-462

[19] Kim SJ, Lee SC, Hong SK, An K, An G, Kim SR. Ectopic expression of a cold responsive OsAsr1 cDNA gives enhanced cold tolerance in transgenic rice plants. Molecules and Cells. 2009;**27**:449-458

[20] Tang SY, Li YR, Yang LT. Evaluation of cold tolerance and photosynthetic characteristics in different sugarcane genotypes. Journal of Global Bioscience. 2015;**4**:2459-2467

[21] Garstka M, Venema JH, Rumak I, Gieczewska K, Rosiak M, KoziolLipinska J, et al. Contrasting effect of dark-chilling on chloroplast structure and arrangement of chlorophyll–protein complexes in pea and tomato: Plants with a different susceptibility to non-freezing temperature. Planta. 2007;**226**:1165-1181

[22] Strauss AJ, Kruger GHJ, Strasser RJ, VanHerden PDR. The role of low soil temperature in the inhibition of growth and PSII function during dark chilling in soybean genotypes of contrasting tolerance. Physiologia Plantarum. 2007;**131**:89-105

[23] Kim SI, Kim DM, Tai TH. Evaluation of rice seedling tolerance to constant and intermittent low temperature stress. Rice Science. 2012;**19**:295-308

[24] Passarini F, Wientjes E, Hienerwadel R, Croce R. Molecular basis of light harvesting and photoprotection in CP24: Unique features of the most recent antenna complex. The Journal of Biological Chemistry. 2009;**284**:29536-29546

[25] Han H, Gao S, Li B, Dong XC, Feng HL, Meng QW. Overexpression of violaxanthin de-epoxidase gene alleviates photoinhibition of PSII and PSI in tomato during high light and chilling stress. Journal of Plant Physiology. 2010;**167**:176-183

[26] Laugier E, Tarrago L, Vieira Dos Santos C, Eymery F, Havaux M, Rey P. Arabidopsis thaliana plastidial methionine sulfoxidereductases B, MSRBs, account for most leaf peptide MSR activity and are essential for growth under environmental constraints through a role in the preservation of photosystem antennae. The Plant Journal. 2010;**61**:271-282

[27] Anjum NA, Ahamd I, Mohmood I, Pacheco M, Duarte AC, Pereira E. Modulation of glutathione and its related enzymes in plants responses to toxic metals and metalloids-a review. Environmental and Experimental Botany. 2012;**75**:307-324

[28] Gill T. Reactive oxygen species and antioxidant machinery in abiotic stress tolerance in crop plants. Plant Physiology and Biochemistry. 2010;**48**:909-930

[29] Miller G, Suzuki N, Ciftci-Yilmaz S, Mittler R. Reactive oxygen species homeostasis and signalling during drought and salinity stresses. Plant, Cell & Environment. *2010*;**33**:453-467

[30] Sharma P, Jha AB, Dubey RS, Pessarakli M. Reactive oxygen species,

oxidative damage and antioxidative defense mechanism in plants under stressful conditions. Journal of Botany. 2012;**12**:217037

[31] Kusvuran S, Kiran S, Ellialtioglu SS. Antioxidant Enzyme Activities and Abiotic Stress Tolerance Relationship in Vegetable Crops. Abiotic and Biotic Stress in Plants - Recent Advances and Future Perspectives. InTech; 2016. pp. 490-492. DOI: 10.5772/62235

[32] Mo Y, Liang G, Shi W, Xie J. Metabolic responses of alfalfa (Medicago Sativa L.) leaves to low and high temperature induced stresses. African Journal of Biotechnology. 2010;**10**:1117-1124

[33] Zhu JH, Dong CH, Zhu JK. Interplay between cold-responsive gene regulation, metabolism and RNA processing during plant cold acclimation. Current Opinion in Plant Biology. 2007;**10**:290-295

[34] Nazari M, Maali Amiri R, Mehraban FH, Khaneghah HZ. Change in antioxidant responses against oxidative damage in black chickpea following cold acclimation. Russian Journal of Plant Physiology. 2012;**59**:183-189

[35] Oktem HA, Eyidooan F, Demirba D, Bayrac AT, Oz MT, Ozgur E, et al. Antioxidant responses of lentil to cold and drought stress. Journal of Plant Biochemical Biotechnology. 2008;**17**:15-21

[36] Zhai FF, Han L, Liu JX, Qian YQ, Ju GS, Li W, et al. Assessing cold resistance of mutagenic strains of perennial rye grass under artificial low-temperature stress. Acta Prataculturae Sinica. 2013;**22**:268-279

[37] Khaledian Y, Maali-Amiri R, Talei A. Phenylpropanoid and antioxidant changes in chickpea plants during cold stress. Russian Journal of Plant Physiology. 2015;**62**:772-778

[38] Xie G, Kato H, Sasaki K, Imai R. A cold-induced thioredoxin of rice, OsTrx23, negatively regulates kinase activities of OsMPK3 and OsMPK6 in vitro. FEBS Letters. 2009;**583**:2734-2738

[39] Kader DZA, Saleh AAH, Elmeleigy SA, Dosoky NS. Chilling-induced oxidative stress and polyamines regulatory role in two wheat varieties. Journal of Taibah University Science. 2011;**5**:14-24

[40] Apostolova P, Yordanova R, Popova L. Response of antioxidative defence system to low temperature stress in two wheat cultivars. General and Applied Plant Physiology. 2008;**34**:281-294

[41] Janmohammadi M. Metabolomic analysis of low temperature responses in plants. Current Opinion in Agriculture. 2012;**1**:1-6

[42] Liu W, Yu K, He T, Li F, Zhang D, Liu J. The low temperature induced physiological responses of Avenanuda L., a cold-tolerant plant species. Scientific World Journal. 2013;**2013**:1-7

[43] Takahashi D, Li B, Nakayama T, Kawamura Y, Uemura M. Plant plasma membrane proteomics for improving cold tolerance. Front. Plant Science. 2013;**4**:90

[44] Matteucci M, Angeli SD, Errico S, Lamanna R, Perrotta G, Altamura MM. Cold affects the transcription of fatty acid desaturases and oil quality in the fruit of Oleaeuropaea L. genotypes with different cold hardiness. Journal of Experimental Botany. 2011;**62**:3403-3420

[45] Jiang A, Guo YJ, Fan Y, Xiang BJ, He W, Wang L, et al. Effect of low temperature stress on cold resistance of

Zea mexicana seedling. Pratacultural Science. 2010;**27**:89-92

[46] Li B, Takahashi D, Kawamura Y, Uemura M. Comparison of plasma membrane proteomic changes of arabidopsis suspension-cultured cells (T87 Line) after cold and ABA treatment in association with freezing tolerance development. Plant & Cell Physiology. 2012;**53**:543-554

[47] Suzuki N, Koussevitzky S, Mittler R, Miller G. ROS and redox signaling in the response of plants to abiotic stress. Plant, Cell & Environment. 2011;**35**:259-270

[48] Das K, Roychoudhury A. Reactive oxygen species (ROS) and response of antioxidants as ROS-scavengers during environmental stress in plants. Frontiers in Environmental Science. 2014;**2**:1-13

[49] Azooz MM, Ismail AM, Elhamd MF. Growth, lipid peroxidation and antioxidant enzyme activities as a selection criterion for salt tolerance of maize cultivars grown under salinity stress. International Journal of Agricultural Biology Engineering. 2009;**11**:21-26

[50] Bian S, Jiang Y. Reactive oxygen species, antioxidant enzyme activities and gene expression patterns in leaves and roots of Kentucky bluegrass in response to drought stress and recovery. Science Horticulture. 2009;**120**:264-270

[51] Mohammadian MA, Largani ZK, Sajedi RH. Quantitative and qualitative comparison of antioxidant activity in the flavedo tissue of three cultivars of citrus under cold stress. Australian Journal of Crop Science. 2012;**6**:402-406

[52] Moieni-Korbekandi Z, Karimzadeh G, Sharifi M. Evaluation of total soluble protein and antioxidant activities in two spring cultivars of canola (*Brassica napus* L.) in response to low temperature. International Journal of Agricultural Crop Science. 2013;**5**:401-409

[53] Fan J, Ren J, Zhu W, Amombo E, Fu J, Chen L. Antioxidant responses and gene expression in bermudagrass under cold stress. Journal of the American Society for Horticultural Science. 2014;**139**:699-705

[54] Zhao DY, Shen L, Yu MM, Zheng Y, Sheng JP. Relationship between activities of antioxidant enzymes and cold tolerance of postharvest tomato fruits. Food Science. 2009;**14**:309-313

[55] Wenying L, Yu K, He T, Li F, Zhang D, Liu J. The low temperature induced physiological responses of Avena nuda L., a cold-tolerant plant species. Scientific World Journal. 2013;**20**:13-17

[56] Fahimirad S, Karimzadeh G, Ghanati F. Cold-induced changes of antioxidant enzymes activity and lipid peroxidation in two canola (*Brassica napus* L.) cultivars. Journal of Plant Physiological Breeding. 2013;**3**:1-11

[57] Dai F, Huang Y, Zhou M, Zhang G. The influence of cold acclimation on antioxidative enzymes and antioxidants in sensitive and tolerant barley cultivars. Biologia Plantarum. 2009;**53**:257-262

[58] Aydin S, Buyuk I, Aras S. Relationships among lipid peroxidation, SOD enzyme activity and SOD gene expression profile in Lycopersicum esculentum L. exposed to cold stress. Genetics and Molecular Research. 2013;**12**:3220-3229

[59] Zhang GL, Chen LY, Zhang ST, Zhang H, Liu GH. Effects of high temperature stress on microscopic and ultrastructure characteristics of mesophyll cells in flag leaves of rice. Rice Science. 2009;**16**:65-71

[60] Sun B, Liu GL, Phan TT, Yang LT, Xing YX, Li YR. Effects of cold stress on root growth and physiological metabolisms in seedlings of different sugarcane varieties. Sugar Technology. 2017;**19**:165-175

[61] Javadian N, Karimzadeh G, Mahfoozi S, Ghanati F. Cold-induced changes of enzymes, proline, carbohydrates, and chlorophyll in wheat. Russian Journal of Plant Physiology. 2010;**57**:540-547

[62] Luo Y, Tang H, Zhang Y. Production of reactive oxygen species and antioxidant metabolism about strawberry leaves to low temperature. The Journal of Agricultural Science. 2011;**3**:89-96

[63] Radyuk MS, Domanskaya IN, Shcherbakov RA, Shalygo NV. Effect of low above-zero temperature on the content of low-molecular antioxidants and activities of antioxidant enzymes in green barley leaves. Russian Journal of Plant Physiology. 2010;**56**:175-180

[64] Hajiboland R, Habibi G. Contrastive responses of spring and winter wheat cultivars to chilling and acclimation treatments. Acta Agricultural Sloven. 2011;**97**:233-239

[65] Huang Z, He J, Xia D, Zhong XJ, Li X, Sun LX, et al. Evaluation of physiological responses and tolerance to low-temperature stress of four Iceland poppy (Papaver nudicaule) varieties. Journal of Plant Interactions. 2016;**11**:117-123

[66] Morsy MR, Jouve L, Hausman JF, Hoffmann L, Stewart JMcD. Alteration of oxidative and carbohydrate metabolism under abiotic stress in two rice (*Oryza sativa* L.) genotypes contrasting in chilling tolerance. Journal of Plant Physiology. 2007;**164**:157-167

[67] Zhang WP, Jiang B, Lou LN, Lu MH, Yang M, Chen JF. Impact of salicylic acid on the antioxidant enzyme system and hydrogen peroxide production in Cucumis sativus under chilling stress. Zeitschrift für Naturforschung. Section C. 2011;**66**:413-422

[68] Duan M, Feng HL, Wang LY, Li D, Meng QW. Overexpression of thylakoidal ascorbate peroxidase shows enhanced resistance to chilling stress in tomato. Journal of Plant Physiology. 2012;**169**:867-877

[69] Gong M, Li HY, Li CG. Short Term Cold Shock at 1°C Induced Chilling Tolerance in Maize Seedlings. Vol. 1. Singapore: IACSIT Press; 2011. pp. 346-349

[70] Anjum NA, Umar S, Chan MT. Ascorbate-Glutathione Pathway and Stress Tolerance in Plants. Dordrecht: Springer; 2010. DOI: 10.1007/978-90-481-9404-9

[71] Noctor G, Mhamdi A, Chaouch S, Han YI, Neukermans J, Marquez-Garcia BELEN. Glutathione in plants: An integrated overview. Plant, Cell & Environment. 2012;**35**:454-484

[72] Xu PL, Guo YK, Bai JG, Shang L, Wang XJ. Effects of long-term chilling on ultrastructure and antioxidant activity in leaves of two cucumber cultivars under low light. Physiologia Plantarum. 2008;**132**:467-478

[73] Kumar S, Kaur G, Nayyar H. Exogenous application of abscisic acid improves cold tolerance in chickpea (*Cicer arietinum* L). Journal of Agronomy and Crop Science. 2008;**194**:449-456

[74] Akram NA, Shafiq F, Ashraf M. Ascorbic acid—A potential oxidant scavenger and its role in plant development and abiotic stress tolerance. Frontiers in Plant Science. 2017;**8**:1-17

[75] Zechmann B. Subcellular distribution of ascorbate in plants. Plant Signaling & Behavior. 2011;**6**:360-363

[76] Venkatesh J, Park SW. Plastid genetic engineering in Solanaceae. Protoplasma. 2012;**249**:981-999

[77] Lukatkin AS, Anjum NA. Control of cucumber (*Cucumis sativus* L.) tolerance to chilling stress evaluating the role of ascorbic acid and glutathione. *Environmental Sciences*. 2014;**2**:1-6

[78] Airaki M, Leterrier M, Mateos RM, Valderrama R, Chaki M, Barroso J. Metabolism of reactive oxygen species and reactive nitrogen species in pepper (*Capsicum annuum* L.) plants under low temperature stress. Plant, Cell & Environment. 2012;**35**:281-295

[79] Kumar S, Malik J, Thakur P, Kaishta S, Sharma KD, Upadhyaya HD. Growth and metabolic responses of contrasting chickpea (Cicer arietinum L.) genotypes to chilling stress at reproductive phase. Acta Physiologiae Plantarum. 2011;**33**:779-787

[80] Kim S, Tai TH. Evaluation of seedling cold tolerance in rice cultivars: A comparison of visual ratings and quantitative indicators of physiological changes. Euphytica. 2011;**178**:437-447

[81] Zhang C, Liu J, Zhang Y, Cai X, Gong P, Zhang J. Overexpression of SlGMEs leads to ascorbate accumulation with enhanced oxidative stress, cold, and salt tolerance in tomato. Plant Cell Reports. 2011;**30**:389-398

[82] Esra K, Cemil D, Sulun AU. Effect of cold on protein, proline, phenolic compounds and chlorophyll content of two pepper (*Capsicum annuum* L.) varieties. J Sci. 2010;**23**:1-6

[83] Yadegari LZ, Heidari R, Carapetian J. The influence of cold acclimation on proline, malondialdehyde (MDA), total protein and pigments contents in soybean (Glycine max) seedlings. Journal of Biological Sciences. 2007;**7**:1436-1441

[84] Zuther E, Juszczak L, Lee YP, Baier M, Hincha DK. Time-dependent deacclimation after cold acclimation in Arabidopsis thaliana a accessions. Scientific Reports. 2015;**5**:1-9

[85] Zhang BQ, Yang LT, Li YR. Physiological and biochemical characteristics related to cold resistance in sugarcane. Sugar Technology. 2015;**17**:49-58

[86] Krol A, Amarowicz R, Weidner S. The effects of cold stress on the phenolic compounds and antioxidant capacity of grapevine (*Vitis vinifera* L.) leaves. Journal of Plant Physiology. 2015;**189**:97-104

[87] Wada KC, Mizuuchi K, Koshio A, Kaneko K, Mitsui T, Takeno K. Stress enhances the gene expression and enzyme activity of phenylalanine ammonia-lyase and the endogenous content of salicylic acid to induce flowering in pharbitis. Journal of Plant Physiology. 2014;**171**:895-902

[88] Francisca OG, Juan P. The response of phenylalanine ammonia-lyase, polyphenol oxidase and phenols to cold stress in the olive tree (Olea europaea L. cv. Picual). Journal of Science and Food Agriculture. 2009;**89**:1565-1573

[89] Tsantili E, Shin Y, Nock JF, Watkins CB. Antioxidant concentrations during chilling injury development in peaches. Postharvest Biology and Technology. 2010;**57**:27-34

[90] Toivonen PMA, Hodges DM. Abiotic stress in harvested fruits and vegetables. In: Shanker A, editor. Abiotic Stress in Plants—Mechanisms and

Adaptations. London: InTech; 2011. DOI: 10.5772/22524

[91] Christopoulos TE. Storage of fresh walnuts (Juglans regia L.) -low temperature and phenolic compounds. Postharvest Biology and Technology. 2012;**73**:80-88

[92] Dong CJ, Li L, Shang QM, Liu XY, Zhang ZG. Endogenous salicylic acid accumulation is required for chilling tolerance in cucumber (Cucumis sativus L.) seedlings. Planta. 2014;**240**:687-700

[93] Olenichenko NA, Ossipov V, Zagoskina NV. Effect of cold hardening on the phenolic complex of winter wheat leaves. Russian Journal of Plant Physiology. 2006;**53**:495-500

[94] Pennycooke JC, Cox S, Stushnoff C. Relationship of cold acclimation, total phenolic content and antioxidant capacity with chilling tolerance in petunia (Petunia × hybrida). Environmental and Experimental Botany. 2005;**53**:225-232

[95] Tattini M, Guidi L, Morassi-Bonzi L, Pinelli P, Remorini D, Innocenti E, et al. On the role of flavonoids in the integrated mechanisms of response of Ligustrum vulgare and Phillyrea latifolia to high solar radiation. New Phytologist. 2005;**167**:457-470

[96] Ahmed NU, Park JI, Jung HJ, Hur Y, Nou IS. Anthocyanin biosynthesis for cold and freezing stress tolerance and desirable color in Brassica rapa. Functional & Integrative Genomics. 2015;**15**:383-394

[97] Valenzuela JL, Manzano S, Palma F, Carvajal F, Garrido D, Jamilena M. Oxidative stress associated with chilling injury in immature fruit: Postharvest technological and biotechnological solutions. International Journal of Molecular Sciences. 2017;**18**:2-26

[98] Schulz E, Tohge T, Zuther E, Femie AR, Hincha DK. Flavonoids are determinants of freezing tolerance and cold acclimation in *Arabidopsis thaliana*. Scientific Reports. 2016;**6**:34027

[99] Verbruggen N, Hermans C. Proline accumulation in plants: A review. Amino Acids. 2008;**35**:753-759

[100] Szabados L, Savoure A. Proline: a multifunctional amino acid. Trends in Plant Science. 2010;**15**:89-97

[101] Kaur G, Kumar S, Thakur P, Malik JA, Bhandhari K, Sharma KD, et al. Involvement of proline in response of chickpea (Cicer arietinum L.) to chilling stress at reproductive stage. Scientia Horticulturae. 2011;**128**:174-181

[102] Theocharis A, Bordiec S, Fernandez O, Paquis S, Dhondt-Cordelier S, Baillieul F, et al. Burkholderia phytofirmans strain PsJN primes Vitisvinifera L. and confers a better tolerance to low non-freezing temperatures. Molecular Plant-Microbe Interactions. 2011;**25**:241-249

[103] Hayat S, Hayat Q, Alyemeni MN, Wani AS, Pichtel J, Ahmad A. Role of proline under changing environments: A review. Plant Signaling & Behavior. 2012;**7**:1456-1466

[104] KaviKishor PB, Sreenivasulu N. Is proline accumulation per se correlated with stress tolerance or is proline homoeostasis a more critical issue? Plant, Cell & Environment. 2013;**37**:300-311

[105] Fernandez O, Theocharis A, Bordiec S, Feil R, Jacquens L, Clement C, et al. Burkholderia phytofirmans strain PsJN acclimates grapevine to cold by modulating carbohydrates metabolism.

Molecular Plant-Microbe Interactions. 2012;**25**:496-504

[106] Yuanyuan M, Yali Z, Jiang L, Hongbo S. Roles of plant soluble sugars and their responses to plant cold stress. African Journal of Biotechnology. 2009;**8**:2004-2010

[107] Ruelland E, Vaultier MN, Zachowski A, Hurry V, Kader JC, Delseny M. Cold signalling and cold acclimation in plants. Advances in Botanical Research. 2009;**49**:35-150

[108] Nishizawa A, Yabuta Y, Shigeoka S. Galactinol and raffinose constitute a novel function to protect plants from oxidative damage. Plant Physiology. 2008;**147**:1251-1263

[109] Vanden Ende W, Valluru R. Sucrose, sucrosyl oligosaccharides, and oxidative stress: Scavenging and salvaging? Journal of Experimental Botany. 2009;**60**:9-18

[110] Zeng Y, Yu J, Cang J, Liu L, Mu Y, Wang J, et al. Detection of sugar accumulation and expression levels of correlative key enzymes in winter wheat (*Triticum aestivum*) at low temperatures. Bioscience, Biotechnology, and Biochemistry. 2011;**75**:681-687

[111] Fernandez O, Bethencourt L, Quero A, Sangwan RS, Clement C. Trehalose and plant stress responses: Friend or foe? Trends in Plant Science. 2010;**15**:409-417

[112] Miranda JA, Avonce N, Suarez R, Thevelein JM, Van Dijck P, Iturriaga G. A bifunctional TPS-TPP enzyme from yeast confers tolerance to multiple and extreme abiotic-stress conditions in transgenic Arabidopsis. Planta. 2007;**226**:1411-1421

[113] Dhingra M. Physiological responses and tolerance mechanisms of low temperature stress in plants. International Journal of Advanced Research. 2015;**3**:637-646

[114] Janmohammadi M, Enayati V, Sabaghnia N. Impact of cold acclimation, de-acclimation and re-acclimation on carbohydrate content and antioxidant enzyme activities in spring and winter wheat. Icelandic Agricultural Sciences. 2012;**25**:3-11

[115] Turhan E, Ergin S. Soluble sugars and sucrose-metabolizing enzymes related to cold acclimation of sweet cherry cultivars grafted on different root stocks. Scientific World Journal. 2012;**2012**:1-7

[116] Streb P, Aubert S, Gout E, Feierabend J, Bligny R. Cross tolerance to heavy-metal and cold-induced photoinhibition in leaves of Pisum sativum acclimated to low temperature. Physiology and Molecular Biology of Plants. 2008;**14**:185-193

[117] Partelli FL, Vieira HD, Viana AP, Batista-Santos P, Rodrigues AP, Leitao AE, et al. Low temperature impact on photosynthetic parameters of coffee genotypes. Pesquisa Agropecuária Brasileira. 2009;**44**:1404-1415

[118] Burchett S, Niven S, Fuller MP. Effect of cold acclimation on the water relations and freezing tolerance of Hordeum vulgare L. Cryo Letters. 2006;**27**:295-303

[119] Klotz KL, Haagenson DM. Wounding, anoxia and cold induce sugarbeet sucrose synthase transcriptional changes that are unrelated to protein expression and activity. Journal of Plant Physiology. 2008;**165**:423-434

[120] Abdel-Latif A. Activity of sucrose synthase and acid invertase in wheat seedlings during a cold-shock using

micro plate reader assays. Australian Journal of Basic and Applied Sciences. 2008;**2**:53-56

[121] Jouve L, Hoffmann L, Hausman JF. Polyamine, carbohydrate, and proline content changes during salt stress exposure of aspen (Populustremula L.): Involvement of oxidation and osmoregulation metabolism. Plant Biology. 2004;**6**:74-80

[122] Yue C, Cao HL, Wang L, Zhou YH, Huang YT, Hao XY, et al. Effect of cold acclimation on sugar metabolism and sugar-related gene expression in tea plant during the winter season. Plant Molecular Biology. 2015;**88**:591-608

[123] Cowie IR, Ebshish OS, Mohamed KS, Pearce RS. Sucrose helps regulate cold acclimation of Arabidopsis thaliana. Journal of Experimental Botany. 2008;**15**:4205-4217

[124] Janmohammadi M, Zolla L, Rinalducci S. Low temperature tolerance in plants: Changes at the protein level. Phytochemistry. 2015;**117**:76-89

[125] Griffith M, Yaish MW. Antifreeze proteins in overwintering plants: A tale of two activities. Trends in Plant Science. 2004;**9**:399-405

[126] Griffith M, Lumb C, Wiseman SB, Wisniewski M, Johnson RW, Marangoni AG. Antifreeze proteins modify the freezing process in planta. Plant Physiology. 2005;**138**:330-340

[127] Yaish MW, Doxey AC, McConkey BJ, Moffatt BA, Griffith M. Cold-active winter rye glucanases with ice-binding capacity. Plant Physiology. 2006;**141**:1459-1472

[128] Wang X, Li W, Li M, Welti R. Profiling lipid changes in plant response to low temperature. Physiologia Plantarum. 2006;**126**:90-96

[129] Seo PJ, Kim MJ, Park JY, Kim SY, Jeon J, Lee YH, et al. Cold activation of a plasma membrane-tethered NAC transcription factor induces a pathogen resistance response in Arabidopsis. The Plant Journal. 2010;**61**:661-671

[130] Xu J, Li Y, Sun J, Du L, Zhang Y, Yu Q, et al. Comparative physiological and proteomic response to abrupt low temperature stress between two winter wheat cultivars differing in low temperature tolerance. Plant Biology. 2013;**15**:292-303

[131] Sarhadi E, Mahfoozi S, Hosseini SA, Salekdeh GH. Cold acclimation proteome analysis reveals close link between up-regulation of low-temperature associated proteins and vernalization fulfillment. Journal of Proteome Research. 2010;**9**:5658-5667

[132] Hon WC, Griffith M, Chong P, Yang DSC. Antifreeze proteins in winter rye are similar to pathogenesis-related proteins. Plant Physiology. 1995;**109**:879-889

[133] Chang M, Chen S, Lee S, Chen Y. Cold-acclimation and root temperature protection from chilling injury in chilling-sensitive mung bean (Vigna radiata L.) seedlings. Botany Bulletin Academic Sinica. 2001;**42**:53-60

[134] Gharechahi J, Alizadeh H, Naghavi MR, Sharifi G. A proteomic analysis to identify cold acclimation associated proteins in wild wheat (Triticum urartu L.). Molecular Biology Reports. 2014;**41**:3897-3905

[135] Yadav SK. Cold stress tolerance mechanisms in plants. A review. Agronomical Sustainable Development. 2010;**30**:515-527

[136] Zhao MG, Chen L, Zhang LL, Zhang WH. Nitric reductase dependent nitric oxide production is involved in cold acclimation and freezing tolerance in Arabidopsis. Plant Physiology. 2009;**151**:755-767

[137] Sato Y, Masuta Y, Saito K, Murayama S, Ozawa K. Enhanced chilling tolerance at the booting stage in rice by transgenic overexpression of the ascorbate peroxidase gene, OsAPXa. Plant Cell Reports. 2011;**30**:399-406

[138] Zahedi H, Alahrnadi SMJ. Effects of drought stress on chlorophyll fluorescence parameters, chlorophyll content and grain yield of wheat cultivars. Journal of Biological Sciences. 2007;7:841-847

[139] Bonnecarrere V, Borsani O, Diaz P, Capdevielle F, Blanco P, Monza J. Response to photoxidative stress induced by cold in japonica rice is genotype dependent. Plant Science. 2011;**180**: 726-732

[140] Zhang Y, Tang HR, Luo Y. Variation in antioxidant enzyme activities of two strawberry cultivars with short-term low temperature stress. World Journal of Agricultural Science. 2008;**4**:458-462

[141] McClung CR, Davis SJ. Ambient thermometers in plants: From physiological outputs towards mechanisms of thermal sensing. Current Biology. 2010;**20**:1086-1092

[142] Hossain MA, Hossain MZ, Fujita M. Stress-induced changes of methylglyoxal level and glyoxalase I activity in pumpkin seedlings and cDNA cloning of glyoxalase I gene. Australian Journal of Crop Science. 2009;**3**:53-64

[143] Janska A, Aprile A, Zamecnik J, Cattivelli L, Ovesna J. Transcriptional responses of winter barley to cold indicate nucleosome remodelling as a specific feature of crown tissues. Functional & Integrative Genomics. 2011;**11**:307-325

[144] Winfield MO, Lu C, Wilson ID, Coghill JA, Edwards KJ. Plant responses to cold: Transcriptome analysis of wheat. Plant Biotech. 2010;**8**:749-777

[145] Hossain MA, Mostofa MG, Fujita M. Cross protection by cold-shock to salinity and drought stress-induced oxidative stress in mustard (*Brassica campestris* L.) seedlings. Mol. Plant Breeding. 2013;**4**:50-70

[146] Li SL, Xia YZ, Liu J. Effects of cold-shock on tomato seedlings under high-temperature stress. Chinese Journal of Applied Ecology. 2014;**25**:2927-2934

[147] Sasaki K, Imai R. Pleiotropic roles of cold shock domain proteins in plants. Frontiers in Plant Science. 2012;**2**:116

[148] Awasthi R, Bhandari K, Nayyar H. Temperature stress and redox homeostasis in agricultural crops. Frontiers in Environmental Science. 2015;**3**:11

[149] Hossain MA, Li ZG, Hoque TS, Burritt DJ, Fujita M, Munne-Bosch S. Heat or cold priming-induced cross-tolerance to abiotic stresses in plants: Key regulators and possible mechanisms. Protoplasma. 2017;**2017**:1-14

[150] Umezawa T, Nakashima K, Miyakawa T, Kuromori T, Tanokura M, Shinozaki K. Molecular basis of the core regulatory network in ABA responses: Sensing, signaling and transport. Plant & Cell Physiology. 2010;**51**:1821-1839

[151] Xu L, Zhang M, Zhang X, Han LB. Cold acclimation treatment induced changes in abscisic acid, cytokinin, and antioxidant metabolism in Zoysiagrass

(Zoysia japonica). Horticultural Science. 2015;**5**:107-180

[152] Pompeiano A, Vita F, Miele S, Guglielminetti L. Freeze tolerance and physiological changes during cold acclimation of giant reed (Arundo donax L.). Grass and Forage Science. 2013;**70**:168-175

[153] Heidarvand L, Amiri RM. What happens in plant molecular responses to cold stress? Acta Physiologiae Plantarum. 2010;**32**:419-431

[154] Conde A, Chaves MM, Geros H. Membrane transport, sensing and signaling in plant adaptation to environmental stress. Plant & Cell Physiology. 2011;**52**:1583-1602

[155] Minami A, Fujiwara M, Furuto A, Fukao Y, Yamashita T, Kamo M. Alterations in detergent-resistant plasma membrane microdomains in Arabidopsis thaliana during cold acclimation. Plant & Cell Physiology. 2009;**50**:341-359

[156] Doherty CJ, Van Buskirk HA, Myers SJ, Thomashow MF. Roles for Arabidopsis CAMTA transcription factors in cold regulated gene expression and freezing tolerance. The Plant Cell. 2009;**21**:972-984

[157] Thomashow MF. Molecular basis of plant cold acclimation: Insights gained from studying the CBF cold response pathway. Plant Physiology. 2010;**154**:571-577

[158] Chen TH, Murata N. Glycinebetaine: An effective protectant against abiotic stress in plants. Trends in Plant Science. 2008;**13**:499-505

[159] Moffatt B, Ewart V, Eastman A. Cold comfort: Plant antifreeze proteins. Physiologia Plantarum. 2006;**126**:5-16

[160] Kargiotidou A, Kappas I, Tsaftaris A, Galanopoulou D, Farmaki T. Cold acclimation and low temperature resistance in cotton: Gossypium hirsutum phospholipase D a isoforms are differentially regulated by temperature and light. Journal of Experimental Botany. 2010;**61**:2991-3002

[161] Jan M, Shinwari KI, Shah G, Khan MHU, Ullah S, Hameed A, et al. Consequences of short term low temperature stress on physiological and biochemical aspects of rice (Oryza sativa L.). Science Ağrıculture. 2015;**10**:1-14

[162] Razmi Z, Hamidi R, Pirasteh-Anosheh H. Seed germination and seedling growth of three sorghum (Sorghum bicolor L.) genotypes as affected by low temperatures. International Journal of Farmer Alliance Science. 2013;**2**:851-856

[163] Zonouri M, Javadi T, Ghaderi N, Saba MK. Effect of foliar spraying of ascorbic acid on chlorophyll a, chlorophyll b, total chlorophyll, carotenoids, hydrogen peroxide, leaf temperature and leaf relative water content under drought stress in grapes. Bulletin Environmental Pharmacological Life Science. 2014;**3**:178-184

[164] Gerganova M, Stanoeva D, Popova A, Velitchkova M. Pigment content and oxygen evolution of tomato plants as affected by long term treatment at suboptimal temperature. Comptes Rendus de L'Academie Bulgare des Sciences. 2016;**69**:1429

[165] Zhang YH, Chen LJ, He JL, Qian LS, Wu LQ, Wang RF. Characteristics of chlorophyll fluorescence and antioxidative system in super-hybrid rice and its parental cultivars under chilling stress. Biologia Plantarum. 2010;**54**:164-168

[166] Bertamini M, Zulini L, Muthuchelian K, Nedunchezhian N. Low night temperature effects on photosynthetic performance on two grapevine genotypes. Biologia Plantarum. 2007;**51**:381-385

[167] Mishra A, Heyer AG, Mishra KB. Chlorophyll fluorescence emission can screen cold tolerance of cold acclimated *Arabidopsis thaliana* accessions. Plant Methods. 2014;**10**:1-10

[168] Aaron EW, Ryan MW. Differential cold acclimation ability of Petunia spp. Horticultural Science. 2009;**44**:1219-1222

Chapter 6

Perspective Chapter: Rootstock-Scion Interaction Effect on Improving Salt Tolerance in Fruit Trees

Ahlem Zrig, Safia Belhadj, Taieb Tounekti, Habib Khemira and Salma Yousif Sidahmed Elsheikh

Abstract

Salt stress is a sever threat to global agriculture. Improving salt tolerance is a problematic task due to the large number of characteristics involved. Graft technique is a potential substitute to breeding and interesting practices to salt tolerance since it unites a scion and rootstock of two genotypes. Increased salinity tolerance in fruit trees will alter water relations, disrupt ionic balance, which can harm plant tissues and thus limit plant productivity. Therefore, the ability of fruit trees to resist salinity varies by species, although it is mostly determined by the type of their root systems. In this regard, the selection of salt-tolerant rootstocks can help maintain productivity under salinity. Several physiological and biochemical changes are attributed to the favorable response of grafting exerted by tolerant rootstocks or scion-rootstock interactions on yield and fruit attributes of plants in saline environments. Rootstocks provide grafted plants different salt tolerance mechanisms including the accumulation of compatible solutes and enhancing the antioxidant mechanisms in scion. The importance of grafting, strategies for selecting appropriate rootstocks, scion-rootstock interaction for growth and the tolerance mechanisms used by plants to avoid the effects of salt stress, are all discussed in this review. Grafting's potential challenges are also discussed.

Keywords: grafting, rootstocks, scion, salinity, glycophyte

1. Introduction

Environmental stress causes a variety of physiological stress reactions in plants, which can change the chemical composition of crops and, as a result, the quality of harvested goods and linked to the quantity and quality of agricultural seeds and food. The most significant of all environmental stresses is salinity, which affects the plant growth and barricades plants from reaching their genetic potential [1]. Salinity occurs when salts accumulate excessively in the water [2]. High salinity affects about 20% of planted native land and 33% of watered farming lands all around the world, and these regions are growing at a pace of 10% each year. Salinity is expected to affect 50% of all arable land by 2050, according to estimates [3]. Salts in the soil are essential for the growth of plant [3]. However, excess of salt can lead to ion toxicity and osmotic stress and interfere with soil nutrient balance [4] affecting plant growth and physiological functions. Salt stress impairs a plant's capacity for photosynthesis and chlorophyll production, which frequently results in a decrease in PSII activity, an obstruction to electron transport, a restriction on carbon assimilation, and peroxidation or dissociation of the thylakoid membrane [5]. Plant roots are characterized by high developmental plasticity, facilitating adaptation to adverse environments [6]. Because there is less cell development in the root elongation zone, the root responds to salt stress by inhibiting root growth [5]. The most actively developing part of the root tip, the meristematic zone, which is a key component of the architecture of the root system, undergoes modifications as a result of saline stress in order for roots to adapt [7]. Therefore, rootstocks are also commonly employed in current wood fruit crop cultivation because of their capacity to adapt a cultivar to a variety of environmental conditions and cultural methods [8, 9]. Rootstocks, on the other hand, can alter scion performance by decreasing plant vigor and allowing the production of high-density orchards [10]. Several breeding initiatives around the world are working on improving rootstocks for fruit trees. It has been reported that the most important factor limiting the cultivation of fruit trees such as peach, almond, plum, cherry, olive, and grape is their special ecological requirements. Aside from the high lime rate of the soil, various diseases also hamper the cultivation of fruit tree species. As a result, in many Mediterranean regions, rootstock is needed to overcome these limitations while growing stone fruits. Therefore, interspecific hybridization is widely used in rootstock breeding programs to expand the genetic base and allow the introduction of genes not found in the breeding population [11]. Because they are tolerant of lime-induced Fe chlorosis and graft-compatible with peach cultivars, almond peach hybrids are widely used as rootstocks for peach trees in Mediterranean countries [12]. They are robust and suitable for use in poor and dry soils [13]. New varieties resistant to biotic stressors, including root-knot nematodes (Meloidogyne spp.) and replanting conditions, have also been produced [14, 15]. Rootstock/scion interactions are widely known to produce a range of properties that transform the overall performance of the combination and have important economic implications and are referred to as rootstock/scion relationships [16]. However, the mechanisms behind these connections are complicated and poorly understood. Soil salinity affects fruit tree species in particular. For instance, peach [17] and almond tree [18] shoot growth is inhibited by relatively low NaCl concentrations (25 mM) in the soil solution. Stress-tolerant crops must be produced through the selection and breeding of cultivars capable of providing commercial yields in saline or drought conditions to reduce the detrimental effects of these stresses [18]. Unfortunately, due to the genetic complexity of abiotic stress tolerance, this is a challenging task [19]. As a result, various cultural opportunities were scrutinized. In this regard, grafting has been used in crops production in the world to improve plant tolerance to abiotic stress by grafting best cultivars (as scion) onto more resistant genotypes (as rootstocks) [20]. Grafting is distinguished as a crucial process to adapt the plant vigor and extend the crops [8, 21]. The most important factor that affects the adaptability of fruit tree to soil stress is the scion-rootstock combination [22]. The combination of rootstock/scion has the potential to produce a plant with characteristics that neither component would have if grown independently. Indeed the rootstocks can provide many characteristics that do not appear in the scion, such as resistance to pests and diseases in the soil [23, 24], improved nutrient uptake, and better tolerance to high saline [25, 26]. The rootstock vigor can also improve the whole plant vigor, increase fruit quality, and alter fruit ripening period [27]. Grafting can also lengthen the harvesting season [13] and extend the life of the crop after harvest [28]. Based on recent studies in this subject, in this article, we will give an overview of the potential of the grafting method to improve the salt tolerance of fruit trees. The physiological and biochemical elements of the rootstock–scion relationship in fruit trees will be explored, as well as the mechanisms involved in the salt tolerance of grafted plants and the influence of the rootstock on the main physiological and biochemical processes of the scion, in order to boost the tolerance of trees grown under saline conditions.

2. The effect of salinity on fruit trees species

Soil salinity is a global issue endangering land productivity and one of the most important environmental concerns limiting plant production in arid and semiarid areas [16]. Salinity alters water relations, disrupt ionic balance, and produces reactive oxygen species (ROS) that can harm plant tissues and thus limit plant productivity. Salinity also causes osmotic and oxidative stress in plant tissues [29, 30] and affects growth, development [31], and the rate of plant less survival [32, 33].

The most sensitive horticultural crops to high salinity were fruit trees [34], Fruit crops are somewhat affected by salinity levels, and once salt concentrations exceed the threshold, their growth rate drops faster than most crops [35]. In fact, salt stress disrupts many metabolic activities in most glycophyte plants, impairing vital cellular function. Many fruit trees, on the other hand, have evolved systems to withstand salt stress and can thus thrive in saline soils [36]. The ability of fruit trees to resist salinity varies by species, although it is mostly determined by the type of their root systems [22]. The ability of most glycophyte plants, such as fruit trees, to exclude or retain potentially harmful ions, determines their salt tolerance. Different species or root types have different adaptation methods for salt stress. The overall concentration of soluble salts in the soil solution, or its osmotic potential, is linked to reduction in growth and yield [32]. For example, shoot growth of peach [37] and almond [5] is affected by low doses (25 mM NaCl) of salt in the soil solution. Zrig et al. [38] revealed that the sweet almond tree cannot tolerate a concentration higher than 75 mM NaCl. Under this lethal concentration, the total dry weight and shoot extension of almond cv Mazzetto decreased while the root/shoot ratio increased. The ratio root/shoot was a state of homeostasis and is independent of root salinity. In this regard, at high salinity level, many plants allow the big part of biomass to roots to maintain a better uptake of water and therefore reach to optimum autoregulation. In pistachio species (P. *atlantica* and Pistacia lentiscus), the growth, the number of leaves, and the fresh biomass of a 3-month-old plantlets decreased significantly with 10 gL^{-1} of NaCl. The leaf browning rate also increased [34]. Furthermore, several studies have been conducted to determine the effects of varying salt levels on the yield of mature plum trees (*Prunus salicina*, cv Santa Rosa) [39]. At 4 dm/S, plum vegetative development appeared more sensitive, and the same treatment caused significant leaf damage [35]. Excessive doses of salt in soil reduce significantly the performance of fast-growing plants of *Prunus* species [40]. Salt load control systems at the whole plant level strongly integrate growth rates and plant shape [41]. Furthermore, high salinity disturbs the osmotic adjustment and leaf water relations [42, 43]. Stone fruit crops (*Prunus* spp., for instance) such as apricot, cherry, nectarine, almond, peach, and plum are the world's seventh-largest crop producers. Stone fruit trees are sensitive to salinity, particularly to chloride, and irrigation with saline water can significantly reduce yields [44, 45]. The main ions causing problems are Cl^-, SO_4^{2-}, HCO_3^-, Na^+, Ca^{2+}, Mg^{2+}, NO_3^-, and K^+, although in several fields B is concerned [46]. In general, during the seed germination stage, the plants were moderately salt-tolerant, but during the young seedling stage, they become more sensitive and progressively more tolerant as they get older through the reproductive stage [47]. Furthermore, most stone fruit trees are sensitive to salt stressors, and their production gradually decreases at salt concentrations above 1.5 dSm^{-1}, reaching 50% of normal production at 4 dSm^{-1} [48].

Plants are affected by salt stress in various ways, including ionic diseases, osmotic stress, and nutritional imbalances. Overproduction of reactive oxygen species (ROS) such as singlet oxygen (O_2), superoxide anion (O_2^-), hydrogen peroxide (H_2O_2), and hydroxyl radical (OH) is a common feature of these effects, which are highly reactive and toxic and cause damage to proteins, lipids, carbohydrates, and DNA, leading to oxidative stress [49]. Salt stress closes stomata, lowering the CO_2/O_2 ratio inside leaf tissues and preventing CO_2 fixation. Plant antioxidant enzymes work together to prevent uncontrolled oxidation cascades and protect plant cells from oxidative damage by scavenging ROS [50].

The processes of salinity sensing and signal transduction within the plant are not well understood, which is exacerbated by the restricted investigation done on this topic for fruit trees [51]. According to several authors [52, 53], plants respond to high levels of salt by activating a signaling network and a comprehensive reaction that involves the synthesis of a variety of chemicals that maintain cellular homeostasis. When the plants sense high levels of salt, primary and secondary metabolites will be activated to provide adequate osmotic balance. This is the cleanest approach for plants to adapt to salt stress. In almond leaves, when the electrical conductivity of irrigation water exceeds 3 dS m^{-1}, Ranjbarfordoei et al. [44] found that the total chlorophyll content and fluorescence parameters are negatively affected. *In vitro* cultivated "Gisela 6" cherry and exposed to various salt concentrations also showed increased malondialdehyde content as well as expression of SOD, ascorbate peroxidase, catalase, and glutathione reductase. Milosevic et Milošević [54] found that the Myrobalan plum (*Prunus cerasifera* Ehrh.) is one of the most tolerant to salinity and B (reduced salt uptake in the root), while the peach "Nemared" (*Prunus persica* (L.) Batsch) is one of the most sensitive (high salt content in the stem).

Traditional breeding programs have attempted to improve crop salt tolerance, but with limited success due to the trait's complexity [55]. The adoption of resistant rootstocks is a technique to avoid or reduce production losses caused by salinity. In this regard, scion/rootstock pairings with a modest growth response have been demonstrated to have a greater ability to exclude Na^+ and Cl^- from the shoot [56].

3. Effect of the rootstock on the salt tolerance of grafted plants

Plant development and structure can be influenced by the complex interactions between scion and rootstock. Rootstocks provide grafted plants with more favorable tolerance against several environmental stresses (both abiotic and biotic); therefore, one of the most important steps in starting a new plantation is choosing a good rootstock, achieving outstanding tree performance in various ecological zones. In fruit trees, rootstocks have been used to boost fruit yield, flavor, and quality while improving nutrient uptake, transport, and resistance to a variety of environmental challenges. The rootstocks had an impact on the physiological characteristics of the grafted plants, as well as other elements of growth and development [49]. Therefore, using grafting techniques to choose a salt-tolerant root system as rootstock can support plants become more salt-tolerant (**Table 1**). The root system of the rootstock replaces the scions during grafting. If the rootstock is chosen correctly, this method will improve nutrition and water uptake, increase carbon and nitrogen metabolism, and increase the salt tolerance of plants [63, 64]. Depending on the intrinsic qualities of the scion, rootstock, and their functional linkages, as well as the severity of salt stress, the persuasive response of grafting on plant growth and production characteristics may differ under salty environments. A number of studies have suggested that grafting can help plants cope with the negative effects of salt stress. Therefore, the rootstock's involvement in determining the tree's performance under saline conditions is essential [65]. Physical characteristics of the root system, such as lateral and vertical development, lead to increased or reduced uptake of water and minerals, which is one of the reasons for the widespread use of rootstocks to overcome salinity. In order to increase detection of better sources of tolerance in the rootstock selection process, it is crucial to determine which type of test offers the best response in a commercial orchard (**Figure 1**).

3.1 Growth and yield

Rootstocks alter the size and shape of trees by reducing internodes, altering branch angles, and altering dates and rates of active growth [49]. The rootstock's response to salinity was related to improved growth under saline circumstances. Several fascinating research studies on Prunus rootstocks, for example, have indicated that via grafting, the rootstock's degree of tolerance is transmitted to the scion variety. In fact, several studies have shown that grafting onto salt-tolerant rootstock can reduce more effectively the negative effects of salt on scion growth. For example, shoot extension and leaf dry weight of almond cv. Mazzetto grafted on Garnem (GN15), which considered as more salt tolerance, were found to be higher to those grafted on GF677 [43]. Accordingly, it appears that grafting onto specific rootstocks can successfully

Species	Rootstocks genotypes	References
Almond	GN15 (*Prunus dulcis* (Mill.) D.A. Webb × (*Prunus persica* (L.)	[57]
Pistachio	*Akbari, 'Ahmad-Aghai' and 'UCB-1*	[34]
Plum	'Mariana 2624 and Adesoto 101'	[58]
Cherry	CAB 6P' (*Prunus cerasus* L.) and Colt' (*Prunus avium* (L.) L. × *Prunus pseudocerasus* Lindl.)	[59]
Apricots	*Prunus microcarpa*	[60]
Olive	Arvanitolia and Lefkolia	[61]
Citrus	Shaker and Cleopatra mandarin	[62]

Table 1.
Different genotypes of rootstocks of different trees. They are used as most salt-tolerant rootstocks.

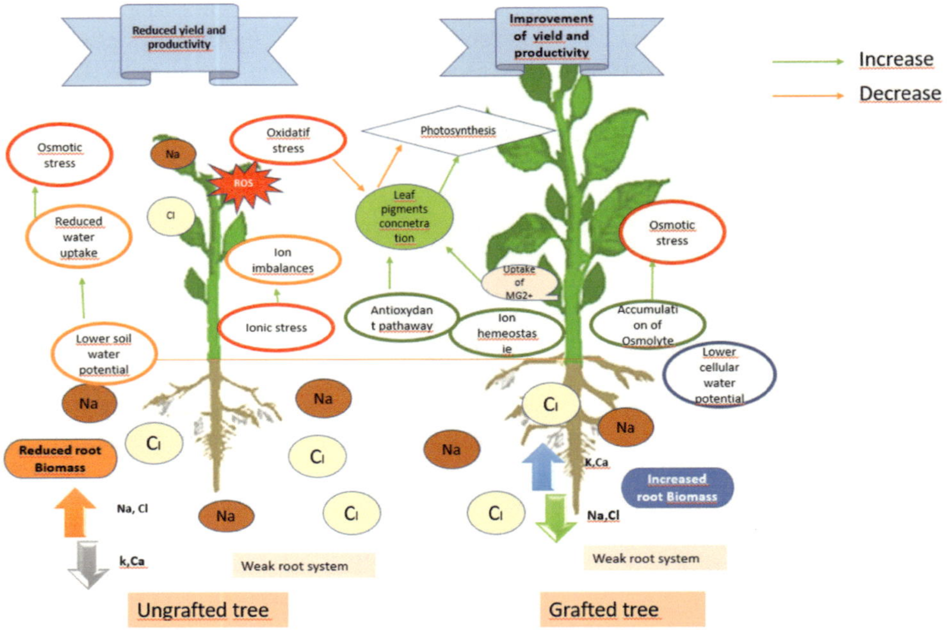

Figure 1.
Potential beneficial effects of rootstocks on trees in saline soil conditions, summarized from the literature. Salinity reduces root biomass because it prevents cell proliferation, which lowers the overall biomass yield (left). Grafting improves roots vaporosity by modifying root architecture and increased nutrient uptake (right). In ungrafted tree, Na^+ and Cl^- accumulation generates competition for nutrient uptake and transport. This results in imbalance of the ionic composition of plant and affects plant's physiological traits (left). In grafted tree, rootstock could improve nutrient uptake and maintenance of ionic homeostasis (right). The high uptake of Na^+ and Cl^- caused by an increase in salt content (left). In grafted tree, the accumulation of osmolytes consequently improves plant's water status (right). In ungrafted tree, increasing salinity causes oxidative stress due to imbalance in reactive oxygen species generation and quenching activities of antioxidants (left). In grafted plants, the oxidative stress was reduced under salt stress (right). Salt stress causes a decline in photosynthesis by decreasing chlorophyll concentration (left). The vigor rootstock has positive effect on photosynthesis under salt stress (right).

minimize the negative effects of salt on the scion; this is consistent with previous observations on various plant species [5]. Similarly, Massai et al. [37] found that peach cv. Armaking grafted onto GF677 inhibited whole plant and lateral root growth more than the Arm/Mrs. combination. This indicated that the first combination was more sensitive to NaCl than the second. Aras and Eşitken [66] studied the responses of cherry plants grafted onto two distinct rootstocks (CAB6P, MaxMa 14, and Mazzard Rootstocks) to short-term salt. This study found that 0900 Ziraat grafted onto MaxMa 14 had the largest decreases in rootstock and shoot diameter by 14 and 14.2%, respectively, while 0900/Mazzard had the smallest decreases in rootstock and shoot diameter by 9.2 and 11%, respectively. Cherry (*Prunus cerasus*) rootstocks can considerably alter the scion's tolerance to salinity, according to Ertuka et al. [67]. Kucukymuk et al. [68] confirmed that sweet cherry trees with the 0900/mazzard combination were more vulnerable to salinity increases than plants with the 0900/mahaleb combination. Further research has shown that when "GF-677" and "Mr.S. 2/5" rootstocks were grafted with peach and exposed to varying NaCl concentrations (0–120 mM), the "GF-677" showed improved sensitivity in terms of growth and net CO_2 assimilation [41].

3.2 Root system architecture

The rootstock's root system characteristics are often the most important criterion for increasing salt tolerance in grafted plants; therefore, the most significant factor for enhancing salt tolerance may be a healthy root system [69]. The root is the first organ to deal with salt excess in soil, due to its activities of searching for and moving water, nutrients, and ions, and it can play a key role in defining plant sensitivity or tolerance to salt stress. Under salt stress, a decline in root growth was observed in woody fruit trees with different levels of salt tolerance. Accordingly, the loss was linked to a decrease in root respiration, which is susceptible to high salinity in soils [70]. Root respiration is also a major source of carbohydrates [71], and it is classified into two types: "maintenance respiration," which includes the preservation of existing tissue, and "growth respiration," which involves the production of new tissue [72]. Maintaining a root respiration is important for producing energy and coping with adverse environments like salinity. Because of the significant requirement for ATP to maintain transport activities across concentration gradients, a higher percentage of root respiration was found to be dedicated to scion maintenance in plants under salt stress. For example, a healthy root system produced more cytokinins and utilized xylem sap to transport water to the shoot system, thereby increasing plant development and yield [73]. Furthermore, root hydraulic conductivity can control plant growth by controlling the water supply to the aboveground plant components [67]. The mechanism, however, is still unknown. Because roots are the primary organ exposed to salt stress, salt-induced suppression of root development is extremely clear. Salt stress, for example, slowed root growth in ungrafted plants, whereas the rate of root dry mass loss was lower in grafted Kinnow mandarin [69] and Pistachio [74]. According to recent study, the first defense under salinity takes place in the root system and apoplastic barrier differentiation contribute to salinity response [35]. The exodermis and the endodermis serve as two apoplastic barriers contributing to protection against abiotic stress, such salinity, by regulating the uptake and transport of water and ions from the soil. An anatomical analysis of suberin in almonds roots revealed that the most salt tolerance rootstock exhibits a strinking increase of suberine deposition at both exodermal and endodermal cells under salinity stress [35]. Furthermore, the lignin deposition, particularly at the Casparian strip, is also critical for the regulation of solute transport at apoplastic barriers [75].

3.3 Gas exchange attributes

The first signs of salt stress in glycophytes are the decrease in leaf development, which seems to be driven by a decrease in stomatal conductance [76]. The stomatal closure reduces the assimilation rate of CO_2 in both grafted and non-grafted salt-treated plants by impairing CO_2 diffusion into the leaves as a result of mesophyll conductance impairment. Photosynthesis is blocked by salt stress because it suppresses the electron transport chain leading to photoinhibition, which leads to a decline in plant growth [57]. Earlier findings have demonstrated that grafting onto salt-tolerant rootstocks can enhance photosynthetic efficiency by preserving chloroplast composition and reducing oxidative damage, which ultimately leads to a delay in photoinhibition rate [32]. The photosynthetic rate (A), stomatal conductance (gs), and transpiration of many prunus rootstocks, such as almond, were all impacted by salinity [77]. It is worth noting that the grafting can modify the photosynthetic performance depending on the scion-rootstocks combinations. Several works reported that the grafting on salt-tolerant rootstocks reduced the salt effect on gas exchanges of scion. In fact, research on almond rootstocks found that after 4 weeks of NaCl treatment, A, gs, and rate of transpiration rate (E) were all lower in all grafting combinations compared to control. Mazzetto/GF677 reliably showed the ultimate reductions. By supporting stomatal conductance, the leaves of Mazzetto/Garnem plants were able to lessen the harm of photosynthetic apparatus. In peach trees, Massai et al. [37] showed that the severe reduction in net assimilation rates was observed in Arm/Mrs. compared to Arm/GF during the first 2 weeks of stress imposition. Besides, these reductions in net assimilation rate seemed to be mainly limited by CO2 diffusion into the leaves during the first 15 days of salt stress. A greater decrease in stomatal conductance was obvious in salt-treated Arm/GF leaves, clearly resulting from salt-induced water stress. Goharrizi et al. [29] examined the photosynthetic activity of three Iranian cultivars of pistachio grafted on four rootstocks and found that trees grown on Sarakhs and *P. atlantica* rootstocks had the highest photosynthetic rates. Sharma et al. [78] recorded higher leaf chlorophyll content in Non Pareil almond and bitter almond rootstock combination than on wild peach rootstock, although wild peach rootstock had higher leaf photosynthesis.

3.4 Mineral content

Fruit species, such as prunus or citrus, are categorized as glycophyte plants because they are salinity sensitive on a scale ranging from moderate to severely salt sensitive. Ion transporters' activities are affected by rootstocks, which alter ion intake and transfer to the scion. Rootstocks promote the acquisition of vital elements while reducing salt uptake and transport (e.g., Na^+ and Cl^-) under saline conditions by exclusion or retention of ions. It is thought that effective rootstocks will be able to limit salt ion uptake and their transit to the shoot, hence slowing or preventing hazardous salt ion accumulation in the leaves [79].

The ability of rootstocks to minimize toxicity of Na^+ and/or Cl^- by exclusion and/or reduction of Cl- uptake by the roots, and replacement or substitution of total K^+ by total Na^+ in the foliage, is related to increased salt tolerance by grafting [80, 81]. The rootstocks minimize Na^+ and Cl^- loading and transport to the scion while enhancing K, Ca, and Mg^{2+} ion intake and allowing for low energy osmotic potentials [82]. The salt sensitivity of fruit tree is related to the sensitivity of the leaves to chloride, while the rootstock salt tolerance is related to the rootstock's ability to exclude chloride and protect the scion leaves [26]. The physical characteristics of the root system, such as lateral and vertical development, which results in increased or reduced uptake of water and minerals, have been attributed to the rootstock's influence on the concentration of certain minerals in the aerial parts of the plant; this is one of the reasons for the widespread use of rootstocks to overcome salinity [83]. Other research has shown that increased salt tolerance in grafted plants is related to increased K^+, Ca^{2+}, or Mg^{2+} translocation to the leaves [53]. The rootstock's ability to minimize damaging ion uptake and transport over time is thought to influence shoot growth [84]. According to Zrig et al. (2016) [85], Mazzetto/Garnem had lower Na + content in the aerial sections than Mazzetto/GF677. The greater shoot extension and leaf biomass of Mazzetto/Garnem found in this study can be explained, at least in part, by such an exclusion mechanism. In this regard, scion/rootstock pairings with a modest growth response have been shown to have a greater ability to exclude Na^+ and Cl^- from the shoot. The ability to stock Na^+ and Cl^- in the cell vacuole and excrete salts outside of the leaf cells via specialized organs is a crucial determinant for salt tolerance in glycophytes [86]. Massai et al. [87] showed that a gradient in Na^+ and Cl^- concentration (tissue water molar basis) was observed in leaves of various age of Arm/GF plants, since basal leaves of 120 mM salt-treated plants had a $Na^+ + Cl^-$ concentration 2.1 and 3.4 times higher than that of medial or apical leaves, respectively. On the contrary, the concentration of $Na^+ + Cl^-$ did not differ between leaves of different age of both 80 and 120 mM salt-treated Arm/MrS plants. According to Küçükyumuk et al. [68], when increasing salinity of the irrigation solution, leaves of sweet cherry trees grafted onto mazzard rootstock suffered more than sweet cherry trees grafted onto mahaleb. Na^+ and Cl^- ions can build up in high concentrations in leaves, causing leaf burn in trees grafted with mahaleb.

Rootstocks protect scion shoots from salt damage primarily by reducing ionic stress and, to a lesser extent, by enhancing K^+, Ca^{2+}, and Mg^{2+} translocation to the shoots and leaves, according to studies, but they have minimal role in reducing osmotic stress [88]. Exclusion or restricted uptake by the roots is used to reduce Na^+ and/or Cl^- translocation to the shoot system. According to Pérez-Alfocea et al. [89], plants use their inherent potential to exclude Na^+ and/or Cl^- from shoots by preserving energy consumption in the root system from risky ion efflux.

Aside from osmotic balancing and root exclusion or restricted root-to-shoot transfer of the harmful ion (i.e., Na^+), plants use ion buildup and subsequent partitioning among plant organs or compartmentalization in cellular organs such as vacuoles to reduce salt toxicity. The grafted plants were able to maintain favorable K^+/Na^+, Ca^{2+}/Na^+, and Mg^{2+}/Na^+ ratios in actively growing leaves [77].

By reducing Na^+ toxicity, rootstocks have been proven to restore plant salt tolerance. According to several studies, the increased salt tolerance of grafted plants such as prunus cultivars is due to a decrease in the accumulation of Na + and/or Cl^- in the shoots of the plant [77]. Other research has shown that increased salt tolerance in grafted plants is associated with better transfer of K^+, Ca^{2+}, or Mg^{2+} to the leaves [53]. A 19-year-old commercial Japanese plum (*P. salicina* Lindl. var. salicina) orchard grafted onto "Marianna 2624" rootstock and received varying amounts of a mixture of NaCl and $CaCl_2$ ranging from 0 to 28 mM. This study found that woody tissues accumulated a lot of Na^+ and Cl^-, while the leaves accumulated a lot of Cl^-, leading to leaf lesions. They found that woody tissue can presumably retain the flow of Na^+ to the leaves (which was not observed in young trees). Chlorides have the greatest damaging influence at the leaf level, lowering net photosynthesis, total carbohydrates, chlorophyll content, and leaf area, according to the second part of the invitation [9]. The ability to keep the rate of buildup of harmful ions (Na^+/Cl^-) as close to physiological homeostatic capacity as possible disturbs the plant's response to salinity. Martinez-Rodriguez et al. [84] state that these characteristics can be transmitted to a more salt-sensitive cultivar by using tolerant genotypes as rootstocks [84]. Despite the fact that ionic homeostasis is mandatory for plant persistence, tissue ion concentration is not a reliable measure of salt tolerance [90, 91].

Different amounts of NaCl and $CaCl_2$ mixture from 0 to 28 mM were administered to a 19-year-old commercial orchard of Japanese plum (*P. salicina*) grafted on the rootstock "Marianna 2624" [92]. These researchers found that woody tissues stored a lot of Na^+ and Cl^-, while leaves mainly accumulated Cl^-, leading to leaf lesions. They found also that woody tissue can apparently retain the movement of Na^+ to the leaves (which was not observed in young trees). The same authors observed that the most negative impact was caused by chlorides at leaf level, decreasing net photosynthesis, total carbohydrates, chlorophyll content, and leaf area [92].

3.5 Water relations

Under normal conditions, grafted plants frequently show improved mineral and water uptake compared to non-grafted plants due to improved root vigor of the rootstock [81]. The plasticity in hydraulic properties in a crop species may be offered through increasing resistance along the pathway of water transport in the plant, and this is central to the maintenance of an adequate water supply to the foliage. In glycophyte plants, which include most crops, high salt content in the medium has a well-known influence on plant growth. First, the saline soil's osmotic potential decreased considerably, resulting in a loss in root water uptake ability. Osmotic stress impacts root and leaf growth as well as stomatal conductance and photosynthetic rate. Water relations in the rootstock-scion system have been explored with a focus on increasing plant adaptability to stressful situations. Zrig et al. [38] showed a decline of the relative water content (RWC) similarly in two Mazzetto/GF677 and Mazzetto/GN15 combinations. On the other hand, Massia et al. [37] reported that the greatest reductions in relative growth rate (RGR) observed in salt-treated Arm/GF as compared to Arm/MrS plants. Nevertheless, at the lowest NaCl doses (25 mM), the water potential dropped significantly, especially in Mazzetto/GF677. Compared to GF677, plants grafted onto Garnem consistently had higher leaf water potential, indicating that Mazzetto/Garnem had higher turgor potential. Salinity affected RWC values in 0900/mazzard trees more than 0900/mahaleb trees, according to a study on sweet cherry [68]. Rootstocks give grafted plants a stronger and wider root system, which allows them to absorb more minerals and water than non-grafted plants [57].

4. Some mechanisms of salt tolerance in grafted plants

Salt tolerance mechanisms have been shown to be transported from rootstock to scion [93]. In fact, several mechanisms of salt tolerances have been observed in the rootstock (**Figure 1**), after grafting it has been observed in the scion. While, in some varieties their salt tolerance was different between scion and rootstocks, this may be due to the interaction between the scion and the rootstock through the graft association. Through their roots, plants were evolved different mechanisms of salt tolerance such as exclusion of salts, accumulation of salts for osmotic adjustment, and activation of antioxidative system.

4.1 Accumulation of compatible solutes

To maintain turgor and water intake for growth, plants must keep their internal water potential below that of the soil. This requires an increase in cell osmotic concentration, either by inorganic ion uptake or by the manufacture of metabolically suitable solutes such as sucrose and proline [32]. Osmoregulation is further facilitated by cytosolic compatible solutes, which impede water outflow to the apoplast and vacuole [2]. In this regard, Zrig et al. (2016) [38] showed that Mazzetto/GN15 had higher concentrations of polyamine, proline, and total soluble sugar (TSS) in their leaves than those of Mazzetto/GF677. This highest accumulation of compatible solute was observed in GN15 rootstocks more than on GF677 rootstocks [5, 38]. These results suggest that the rootstock Garnem enhanced the accumulation of those osmolytes in the leaves of the scion (Mazzetto) to be more tolerant to salinity. Still in almond tree, Zrig et al. [38] testified that GN15 rootstocks accumulated anthocyanin in their leaves, and that these polyphenols were reinforced under salt stress. These anthocyanin productions were observed in leaves of Mazzetto grafted onto Garnem (GN15) also under normal conditions; thereafter, the anthocyanin content was sharply increased under a high level of salinity. According to this earlier study, rootstocks induce the anthocyanin biosynthesis in scion leaves to effectively contribute to osmoprotection under salt stress. Thus, the lower salt-tolerant plants Mazzetto grafted onto GF677 rootstocks showed novel production of polyphenols and polyamines to contribute to osmoregulation [43].

4.2 Antioxidant mechanisms

Photosynthesis can be reduced by a reduction in photochemical activity under salt stress, regardless of stomatal conductance [17]. Several abiotic stressors have been shown to decrease Chla+b content and photochemical efficiency of photosystem II (PSII), still these effects can be improved by grafting. In fact, it was demonstrated that the total chlorophyll contents were reduced in the leaves of Mazzetto grafted on GF677; however, it remained unchanged in the plants grafted on Garnem [38]. Accordingly, the rootstock appears to have influenced the rate of Chl turnover or biosynthesis in scion leaves. Similarly, Aras and Eşitken [66] showed that the lesser decrease of cholrophyll pigment is related to the salt tolerance of rootstocks. In fact, in this study on cherry trees, the plants (0900) grafted on Mazzard displayed the lowest decline in chlorophyll content by 7% under salinity compared to those grafted onto CAB 6P and MaxMa 14. Such effect could be due to its ability to maintain higher chlorophyll levels or its ability to block chlorophyllase activity, which causes chlorophyll degradation [66]. Plants also have the ability to scavenge or detoxify ROS produced by salinity to safeguard photosynthesis and prevent the breakdown of cholorophyll pigments. In grafted crops, such antioxidants have been used as indicators for salinity tolerance. Accordingly, an effective antioxidant system is a key determinant of increased salt tolerance of grafted plants. This is accomplished by increasing the activity of antioxidant enzymes and the amount of non-enzymatic antioxidants in the plant. Under NaCl stress, non-enzymatic antioxidant activity was found to be considerably higher in the leaves of grafted eggplant seedlings than in self-rooted seedlings [94]. There has been less research on the antioxidant system in the roots of grafted plants under salt stress than in the leaves. Non-enzymatic antioxidants are higher in sweet almonds grafted on salt-tolerant rootstocks than in sweet almonds grafted onto salt-sensitive rootstocks. Indeed, research on sweet almond cv. Mazzetto found that Mazzetto/Garnem leaves had higher carotenoid/Chl and anthocyanin/Chl ratios than Mazzetto/GF677 foliage. These findings suggest that the light-harvesting antenna and photosystems are protected by more effective antioxidant mechanisms. Carotenoids protect chlorophylls from oxidative damage by scavenging two ROS, singlet molecular oxygen and peroxyl radicals, in addition to their role as light-harvesting pigments that aid in photosynthesis [95]. These results show that Garnem rootstock improved the antioxidation capacity of Mazzetto cells. Mazzetto/Garnem leaves had higher anthocyanin concentrations and anthocyanin/Chl ratios than in Mazzetto/GF677 leaves, even in control plants. The leaves of the scion seem to adopt some of the rootstock's characteristics; for example, the red-leafed Garnem contains more anthocyanins in its tissues than in green-leafed GF677 [38]. The mechanism of this influence is unknown, but growth regulators such as auxin may play a role. Several studies have established the impact of rootstock on the biochemical composition of the scion of several species, including almond [96] and grapevine [63]. The photo-protective importance of anthocyanins against photo-oxidation during salinity stress was highlighted by Zrig et al. [43] in Garnem rootstock, overall antioxidant activity was favorably related to carotenoid and anthocyanin concentrations rather than polyamines. Mazzetto/Garnem plants appear to exhibit the most efficient photoprotective mechanism, which require carotenoids and anthocyanins, as well as more osmoprotectants in the form of proline and soluble sugars. Mazzetto/antioxidant GF677's effect appears to be mostly due to polyphenols and spermidine, both of which were abundant in its leaves.

5. Impact of scion/rootstock reciprocal effects

Rootstock-scion have the capacity to influence tree physiology in several ways, as the relationships between rootstocks and scions are both vast and complex. The scion assimilates the carbohydrates and translocates them, along with the hormones, to the root system. While the root system can also provide hormones for the scion, as well as water and nutrients absorbed from the soil [97]. Numerous research studies have been done to better understand how rootstock and scion interact to produce traits such as modest tree stature and precocity. Further studies have described a main effect of the rootstock and its proficiency to exhibit consistency of growth to the scion, while the scion showed influence in determining tree weight. Physiological research on rootstock-scion interactions revealed a complex relationship that differed depending on the rootstock and scion varieties used [23]. Despite the fact that scion genotype has a dominant effect on most agronomic variables, rootstock-scion-environment interactions introduce new sources of genotypic and phenotypic diversity into the crop. Hydraulic and chemical impulses pass through the xylem and phloem to cause these interactions. Although maximizing crop productivity requires the leaves to obtain an appropriate supply of water and mineral nutrients from the root system via the xylem, integrated plasticity also requires long-distance communication, in which tiny organic molecules play an important role [1]. The xylem and phloem, two plant vascular systems, conduct cross talk between the above and below graft components under normal conditions. Under salt stress, plants accumulate many osmoregulators such as amino acids, carbohydrate, organic acids, and enhance mineral uptake, all these compounds abound in phloem sap, which includes both organic and inorganic components. In a grafted tree, a reciprocal bond between scion and roots develops, in which a progression of reciprocated adjustment develops in the forced connection between two genotypes. Furthermore, the qualities of a scion or rootstock can only be determined in conjunction with the characteristics of other scions/rootstocks. Therefore, the required characteristics of either the scion or the rootstock (e.g., salinity tolerance) should come first when selecting a rootstock-scion combination, followed by a careful selection of a compatible partner [94].

6. Conclusion

The global climate change makes the crop production very vulnerable. In addition, climate change was linked to the increase in temperatures, and fluctuating rainfall patterns cause a significant rise in soil salt, leading to a decline in the production of fruit trees given their sensitivity to this abiotic stressor. In this regard, grafting has been described by several studies as a practice to improve tolerance to a variety of environmental conditions. Indeed, rootstock and scion interaction is effective in

DOI: http://dx.doi.org/10.5772/intechopen.108817

reducing stress factors and producing a healthier plant ideotype. Moreover, the trait of salt tolerance of rootstock was transfer to the aerial part (scion). In fact, the scion could adopt several mechanisms including antioxidant activities to tolerate the salinity. Furthermore, progeny assessment requires an understanding of the mechanisms underlying responses to abiotic stressors. The establishment of a network of assessment sites lets for more information on the performance and salt tolerance mechanism of each rootstock in each season, and which, when combined with the specific tests, leads to a detailed characterization of each genotype, making it easier to select the most appropriate rootstock for each producer's conditions.

Replace the entirety of this text with the main body of your chapter. The body is where the author explains experiments, presents and interprets data of one's research. Authors are free to decide how the main body will be structured. However, you are required to have at least one heading. Please ensure that either British or American English is used consistently in your chapter. Future research should concentrate on understanding the molecular interactions between and within cells that contribute to the salt stress response. With more candidate genes for salinity tolerance being discovered and widely used, genetic engineering has already proven to be an effective method for creating plants that can withstand salty environments. Through interspecific crosses, new sources of variation within the species must be sought. This requires a diverse array of resources that will provide breeders with new options for breaking down resistance and/or tolerance barriers to pests and diseases.

Conflict of interest

"The authors declare no conflict of interest."

Author details

Ahlem Zrig[1,2]*, Safia Belhadj[3], Taieb Tounekti[4,5], Habib Khemira[4]
and Salma Yousif Sidahmed Elsheikh[6]

1 Higher Institute of Preparatory Studies in Biology and Geology, University of Carthage, Tunisia

2 Laboratory of Engineering Processes and Industrial Systems, Chemical Engineering Department, National School of Engineers of Gabes, University of Gabes, Tunisia

3 Ziane Achour University of Djelfa, Route de Moudjebara, Djelfa, Algeria

4 Jazan University, Centre for Environmental Research and Studies, Jazan, Saudi Arabia

5 Faculty of Sciences of Gafsa, University of Gafsa, Campus Universitaire Sidi Ahmed Zarroug, Gafsa, Tunisia

6 College of Science, Department of Biological Science, Northern Boarder University, Arar

*Address all correspondence to: ahlem18zrig@yahoo.fr

References

[1] Schwarz D, Rouphael Y, Colla G, Venema JH. Grafting as a tool to improve tolerance of vegetables to abiotic stresses: Thermal stress, water stress and organic pollutants. Scientia Horticulture (Amsterdam) [Internet]. Elsevier Bacterial vaginosis. 2010;**127**:162-171. Available from: DOI: 10./j.scienta.2010.09.016

[2] Parida AK, Das AB. Salt tolerance and salinity effects on plants: A review. Ecotoxicology and Environmental Safety. 2005;**60**:324-349

[3] Wang CM, Zhang JL, Liu XS, Li Z, Wu GQ, Cai JY, et al. Puccinellia tenuiflora maintains a low Na+ level under salinity by limiting unidirectional Na+ influx resulting in a high selectivity for K+ over Na+. Plant, Cell & Environment. 2009;**32**:486-496

[4] Gupta B, Huang B. Mechanism of salinity tolerance in plants: Physiological, biochemical, and molecular characterization. International Journal of Genomics. 2014;**24**:12-22

[5] Takahashi M, Shigeto J, Sakamoto A, Morikawa H. Selective nitration of PsbO1, PsbO2, and PsbP1 decreases PSII oxygen evolution and photochemical efficiency in intact leaves of Arabidopsis. Plant Signaling and Behavior. [Internet]. Taylor & Francis; 2017;**12**:1-5. Available from:. DOI: 10.1080/15592324.2017.1376157

[6] Liu Z, Bie Z, Huang Y, Zhen A, Niu M, Lei B. Rootstocks improve cucumber photosynthesis through nitrogen metabolism regulation under salt stress. Acta Physiologiae Plantarum. 2013;**35**:2259-2267

[7] Julkowska MM, Koevoets IT, Mol S, Hoefsloot H, Feron R, Tester MA, et al. Genetic components of root architecture remodeling in response to salt stress. Plant Cell. 2017;**29**:3198-3213

[8] Allevato E, Mauro RP, Stazi SR, Marabottini R, Leonardi C, Ierna A, et al. Arsenic accumulation in grafted melon plants: Role of rootstock in modulating root-to-shoot translocation and physiological response. Agronomy. 2019;**9**:2-18

[9] Gainza F, Opazo I, Guajardo V, Meza P, Ortiz M, Pinochet J, et al. Rootstock breeding in prunus species: Ongoing efforts and new challenges. Chilean Journal of Agricultural Research. 2015;**75**:6-16

[10] Rouphael Y, Kyriacou MC. Enhancing quality of fresh vegetables through salinity eustress and biofortification applications facilitated by soilless cultivation. Frontiers in Plant Science. 2018;**9**:1-6

[11] Velasco D, Hough J, Aradhya M, Ross-Ibarra J. Evolutionary genomics of peach and almond domestication: G3 Genes, Genomes. Genetics. 2016;**6**:3985-3993

[12] Socias i Company R, Rubio-Cabetas MJ, Alonso JM, Kodad O, Gómez Aparisi J. An overview of almond cultivars and rootstocks: Challenges and perspectives. Options Méditerranéennes Séries A Mediterrean Seminaire. 2010;**214**:205-214

[13] Baumgartner K, Fujiyoshi P, Ledbetter C, Duncan R, Kluepfel DA. Screening almond rootstocks for sources of resistance to Armillaria root disease. HortScience. 2018;**53**:4-8

[14] Felipe AJ. "Felinem", "Garnem", and "Monegro" almond x peach

hybrid rootstocks. HortScience. 2009;**44**:196-197

[15] Mestre L, Reig G, Betrán JA, Pinochet J, Moreno MÁ. Influence of peach-almond hybrids and plum-based rootstocks on mineral nutrition and yield characteristics of "Big Top" nectarine in replant and heavy-calcareous soil conditions. Horticultural Science. [Internet]. Amsterdam: Elsevier B.V.; 2015;**192**:475-481. DOI: 10.1016/j.scienta.2015.05.020

[16] Bastam N, Baninasab B, Ghobadi C. Improving salt tolerance by exogenous application of salicylic acid in seedlings of pistachio. Plant Growth Regulation. 2013;**69**:275-284

[17] Tattini M, Guidi L, Morassi-Bonzi L, Pinelli P, Remorini D, Degl'Innocenti E, et al. On the role of flavonoids in the integrated mechanisms of response of *Ligustrum vulgare* and *Phillyrea latifolia* to high solar radiation. New Phytologist. 2005;**167**:457-470

[18] Cuartero J, Bolarín MC, Asíns MJ, Moreno V. Increasing salt tolerance in the tomato. Journal of Experimental Botany. 2006;**10**:1045-1058

[19] Iqbal M, Ashraf M. Presowing seed treatment with cytokinins and its effect on growth, photosynthetic rate, ionic levels and yield of two wheat cultivars differing in salt tolerance. 2005;**47**:1315-1325

[20] Öztekin G, Tüzel Y, Gül A, Tüzel IH. Effects of grafting in saline conditions. Acta Horticulturae. 2007;**761**:349-355

[21] Colla G, Rouphael Y, Leonardi C, Bie Z. Role of grafting in vegetable crops grown under saline conditions. Scientia Horticulturae [Internet]. Amsterdam: Elsevier B.V.; 2010;**127**:147-155. DOI: 10.1016/j.scienta.2010.08.004

[22] Alves FO. Rootstock 3 scion interactions on Theobroma cacao resistance to witches' broom: Photosynthetic, nutritional and antioxidant metabolism responses. 2016;**28**:3-10

[23] Gregory PJ, Atkinson CJ, Bengough AG, Else MA, Fernández-Fernández F, Harrison RJ, et al. Contributions of roots and rootstocks to sustainable, intensified crop production. Journal of Experimental Botany. 2013;**64**:1209-1222

[24] Louws FJ, Rivard CL, Kubota C. Grafting fruiting vegetables to manage soilborne pathogens, foliar pathogens, arthropods and weeds. Horticultural Science [Internet]. Amsterdam: Elsevier B.V.; 2010;**127**:127-146. DOI: 10.1016/j.scienta.2010.09.023

[25] Nawaz MA, Imtiaz M, Kong Q, Cheng F, Ahmed W, Huang Y, et al. Grafting: A technique to modify ion accumulation in horticultural crops. Frontiers in Plant Science. 2016;**7**:1-15

[26] Wahb-Allah MA. Effectiveness of grafting for the improvement of salinity and drought tolerance in tomato (*solarium lycopersicon* L.). Asian. Journal of Crop Science. 2014;**25**:112-122

[27] Rouphael Y, Kyriacou MC, Colla G. Vegetable grafting: A toolbox for securing yield stability under multiple stress conditions. Frontiers in Plant Science. 2018;**8**:10-13

[28] Zhao X, Guo Y, Huber DJ, Lee J. Grafting effects on postharvest ripening and quality of 1-methylcyclopropene-treated muskmelon fruit. Horticultural Science [Internet]. Amsterdam: Elsevier B.V.; 2011;**130**:581-587. DOI: 10.1016/j.scienta.2011.08.010

[29] Jamshidi Goharrizi K, Baghizadeh A, Kalantar M, Fatehi F. Combined effects

of salinity and drought on physiological and biochemical characteristics of pistachio rootstocks. Horticultural Science [Internet]. Amsterdam: Elsevier B.V.; 2020;**261**:108970. DOI: 10.1016/j.scienta.2019.108970

[30] Isayenkov SV, FJM M. Plant salinity stress: Many unanswered questions remain. Frontiers in Plant Science. 2019;**10**:30-41

[31] Tavallali V, Karimi S. Methyl jasmonate enhances salt tolerance of almond rootstocks by regulating endogenous phytohormones, antioxidant activity and gas-exchange. The. Journal of Plant Physiology [Internet]. Elsevier; 2019;**234-235**:98-105. DOI: 10.1016/j.jplph.2019.02.001

[32] Munns R. Comparative physiology of salt and water stress. Plant, Cell & Environment. 2002;**25**:239-250

[33] Ouhaddach M, El-Yacoubi H, Douaik A, Hmouni D, Rochdi A. Réponse à la salinité de quelques paramètres physiologiques et biochimiques du Blé (*Triticum aestivum* L.) au stade montaison. Journal of Materials and Environmental. Science. 2016;7:3084-3099

[34] Abidine MKZ, Belhadj Safia LS, Abdelhamid B. Influence DU NACL sur quelques parametres morpho-physiologiques et biochimiques de jeunes plantules de deux influence of nacl on some morpho-physiological and biochemical parameters on young seedlings of two spontaneous species of *Pistacia Genus* P Le. 2020;**10**:2129-2140

[35] Shao Y, Cheng Y, Pang H, Chang M, He F, Wang M, et al. Investigation of salt tolerance mechanisms across a root developmental gradient in Almond rootstocks. Frontiers in Plant Science. 2021;**11**:15-22

[36] Patel MK, Kumar M, Li W, Luo Y, Burritt DJ, Alkan N, et al. Enhancing salt tolerance of plants: From metabolic reprogramming to exogenous chemical treatments and molecular approaches. Cell. 2020;**9**:1-26

[37] Massai R, Remorini D, Tattini M, Plant S, Legnose S, Coltivazioni S, et al. Gas exchange, water relations and osmotic adjustment in two scion/rootstock combinations of Prunus under various salinity concentrations. Stable. 2016;259:153-162. Available from: http://www.jstor.org/stable/24124369 Your use of the JSTOR archive indicates your acceptance of the Te

[38] Zrig A, Ben Mohamed H, Tounekti T, Ennajeh M, Valero D, Khemira H. A comparative study of salt tolerance of three almond rootstocks: Contribution of organic and inorganic solutes to osmotic adjustment. Journal of Agricultural Science and Technology. Tarbiat Modares University; 2015;**17**:675-689

[39] Sidorova T, Miroshnichenko D, Kirov I, Pushin A, Dolgov S. Effect of grafting on viral resistance of non-transgenic plum scion combined with transgenic PPV-resistant rootstock. Frontiers in Plant Science. 2021;**12**:1-14

[40] Toro G, Pimentel P, Salvatierra A. Effective categorization of tolerance to salt stress through clustering prunus rootstocks according to their physiological performances. Horticulturae. 2021;7:12-19

[41] Munns R. Genes and salt tolerance: Bringing them together. New Phytologist. 2005;**25**:645-663

[42] Yang S, Zu Y, Li B, Bi Y, Jia L, He Y, et al. Response and intraspecific differences in nitrogen metabolism of alfalfa (*Medicago sativa* L.) under

cadmium stress. Chemosphere. 2019;**220**:69-76

[43] Zrig A, Ben MH, Tounekti T, Ahmed SO, Khemira H. Differential responses of antioxidant enzymes in salt-stressed almond tree grown under sun and shade conditions. Journal of Plant Science and Research. 2015;**2**:117-127

[44] Ranjbarfordoei A, Samson R, Van Damme P. Chlorophyll fluorescence performance of sweet almond [*Prunus dulcis* (Miller) D. Webb] in response to salinity stress induced by NaCl. Photosynthetica. 2006;**44**:513-522

[45] Zrig A, Ferreira JFS, Hamouda F, Tounekti T, Selim S, Al Jaouni S, et al. The impact of foliar fertilizers on growth and biochemical responses of *Thymus vulgaris* to salinity stress. Arid Land Research and Management[Internet]. Taylor & Francis; 2019;**10**:1-24 Available from: https://www.tandfonline.com/doi/full/10.1080/15324982.2018.1551817

[46] Bernstein N, Meiri A, Zilberstaine M. Root growth of avocado is more sensitive to salinity than shoot growth. The Journal of the American Society for Horticultural Science. 2004;**129**:188-192

[47] Ghada S, Shimaa F. Evaluation of salinity tolerance on new selected almond×peach hybrid rootstocks. Arab Universities Journal of Agricultural Sciences. 2021;**10**:10-20

[48] Dejampour J. Evaluation of salt tolerance in almond [*Prunus dulcis* (L.) Batsch] rootstocks. African Journal Biotechnology. 2012;**11**:11907-11912

[49] Alam A, Ullah H, Attia A, Datta A. Effects of salinity stress on growth, mineral nutrient accumulation and biochemical parameters of seedlings of three citrus rootstocks. International Journal of Fruit Science[Internet]. Taylor & Francis; 2020;**20**:786-804. DOI: 10.1080/15538362.2019.1674762

[50] Talhouni M, Sönmez K, Kiran S, Beyaz R, Yildiz M, Kuşvuran Ş, et al. Comparison of salinity effects on grafted and non-grafted eggplants in terms of ion accumulation, MDA content and antioxidative enzyme activities. Advances in Horticultural Science. 2019;**33**:87-95

[51] Hasegawa PM, Bressan RA, Zhu J-K, Bohnert HJ. Plant cellular and molecular and molecular responses to high salinity. Annual Review of Plant Physiology and Plant Molecular Biology. 2000;**51**:463-499

[52] Shulaev V, Cortes D, Miller G, Mittler R. Metabolomics for plant stress response. Physiology Plant. 2008;**132**:199-208

[53] Zhu JK. Plant salt tolerance. Trends. Plant Science. 2001;**23**:66-71

[54] Milošević N, Milošević T. Seasonal changes and content of sodium in main organs of European plum trees (*Prunus domestica* L.), fruit size and yield as affected by rootstocks on acidic soil. Semina: Ciencias Agrarias. 2012;**33**:605-620

[55] Flowers TJ. Improving crop salt tolerance. Journal of Experimental Botany. 2004;**15**:307-319

[56] Bahrami V, Imani A, Piri S. Evaluation of almond' cultivars growth characteristics under salinity stress. Journal of Novel Applied Sciences. 2015;**87**:1227-1229

[57] Tortosa I, Douthe C, Pou A, Balda P, Hernandez-Montes E, Toro G, et al. Variability in water use efficiency of grapevine tempranillo clones and stability over years at field conditions. Agronomy. 2019;**9**:1-11

[58] Andreu P, Arbeloa A, Lorente P, Marín JA. Early detection of salt stress tolerance of prunus rootstocks by excised root culture. HortScience. 2011;**46**:80-85

[59] Papadakis IE, Veneti G, Chatzissavvidis C, Therios I. Physiological and Growth Responses of Sour Cherry (Prunus cerasus L.) Plants Subjected to Short-Term Salinity Stress. 2018

[60] Larsen SU, Bailly C, Côme D, Corbineau F. Use of the hydrothermal time model to analyse interacting effects of water and temperature on germination of three grass species. Seed Science Research. 2004;**14**:35-50

[61] Asimakopoulou G, Karnaouri A, Staikos S, Stefanidis SD, Kalogiannis KG, Lappas AA, et al. Production of omega-3 fatty acids from the microalga crypthecodinium cohnii by utilizing both pentose and hexose sugars from agricultural residues. Fermentation. 2021;7

[62] Shafieizargar A, Awang Y, Ajamgard F, Juraimi AS, Othman R, Ahmadi AK. Assessing five citrus rootstocks for NaCl salinity tolerance using mineral concentrations, proline and relative water contents as indicators. Asian Journal Plant Science. 2015;**14**:20-26

[63] Sanwal SK, Mann A, Kumar A, Kesh H, Kaur G, Rai AK, et al. Salt tolerant eggplant rootstocks modulate sodium partitioning in tomato scion and improve performance under saline conditions. Agriculture. 2022;**12**

[64] Shahid MA, Balal RM, Khan N, Zotarelli L, Liu G, Ghazanfar MU, et al. Ploidy level of citrus rootstocks affects the carbon and nitrogen metabolism in the leaves of Chromium-stressed Kinnow mandarin plants. Environmental and Experimental Botany[Internet]. Elsevier B.V.; 2018;**149**:70-80. DOI: 10.1016/j.envexpbot.2018.02.010

[65] Grieve C, Grattan S. Mineral nutrient acquisition and response by plants grown in saline environments. Open Journal of Forestry. 1999;**203**:29-35

[66] Aras S, Eşitken A. Responses of cherry plant grafted onto CAB-6P, MaxMa 14 and mazzard rootstocks to short term salinity. Journal of Agricultural Studies. 2019;**7**:29

[67] Erturk U, Sivritepe N, Yerlikaya C, Bor M. Responses of the cherry rootstock to salinity in vitro. Biologia Plantarum. 2007;**10**:2-6

[68] Küçükyumuk C, Yildiz H, Küçükyumuk Z, Ünlükara A. Responses of "0900 Ziraat" sweet cherry variety grafted on different rootstocks to salt stress. Notulae Botanicae Horti Agrobotanici Cluj-Napoca. 2015;**43**:214-221

[69] Sinha AK, AbdElgawad H, Zinta G, Dasan AF, Rasoloniriana R, Asard H, et al. Nutritional status as the key modulator of antioxidant responses induced by high environmental ammonia and salinity stress in European sea bass (*Dicentrarchus labrax*). PLoS One. 2015;**10**:1-29

[70] Jacoby RP, Taylor NL, Millar AH. The role of mitochondrial respiration in salinity tolerance. Trends in Plant Science[Internet]. Elsevier Ltd; 2011;**16**:614-623. DOI: 10.1016/j.tplants.2011.08.002

[71] Rachmilevitch S, Lambers H, Huang B. Root respiratory characteristics associated with plant adaptation to high soil temperature for geothermal and turf-type Agrostis species. Journal of Experimental Botany. 2006;**57**:623-631

[72] Meloni DA, Oliva MA, Martinez CA, Cambraia J. Photosynthesis and activity of superoxide dismutase, peroxidase and glutathione reductase in cotton under salt

stress. Environmental and Experimental Botany. 2003;**49**:69-76

[73] Grace JK, Sharma KL, Seshadri KV, Ranganayakulu C, Subramanyam KV, Raj GB, et al. Evaluation of sweet orange (*Citrus sinensis* L. Osbeck) cv. sathgudi budded on five rootstocks for differential behavior in relation to nutrient utilization in alfisol. Communications in Soil Science and Plant Analysis. 2012;**43**:985-1014

[74] Raoufi A, Rahemi M, Salehi H, Javanshah A. Selecting high performance rootstocks for pistachio cultivars under salinity stress based on their morpho-physiological characteristics. International Journal of Fruit Science[Internet]. Taylor & Francis; 2020;**20**:S29-S47. DOI: 10.1080/15538362.2019.1701616

[75] Naseer S, Lee Y, Lapierre C, Franke R, Nawrath C, Geldner N. Casparian strip diffusion barrier in arabidopsis is made of a lignin polymer without suberin. Proceedings of the National Academy of Sciences. USA. 2012;**109**:10101-10106

[76] Omamt EN, Hammes PS, Robbertse PJ. Differences in salinity tolerance for growth and water-use efficiency in some amaranth (*Amaranthus* spp.) genotypes. New Zealand Journal of Crop and Horticultural Science. 2006;**34**:11-22

[77] Zrig A, Tounekti T, Vadel AM, Ben Mohamed H, Valero D, Serrano M, et al. Possible involvement of polyphenols and polyamines in salt tolerance of almond rootstocks. Plant Physiology and Biochemistry. 2011;**49**:1313-1322

[78] Sharma RM, Dubey AK, Awasthi OP. Physiology of grapefruit (*Citrus paradisi* Macf.) cultivars as affected by rootstock. The Journal of Horticultural Science and Biotechnology. 2015;**90**:325-331

[79] Zrig A, Ben Mohamed H, Tounekti T, Khemira H, Serrano M, Valero D, et al. Effect of rootstock on salinity tolerance of sweet almond (cv. Mazzetto). South African Journal of Botany. Elsevier; 2016;**102**:50-59

[80] Estañ MT, Martinez-Rodriguez MM, Perez-Alfocea F, Flowers TJ, Bolarin MC. Grafting raises the salt tolerance of tomato through limiting the transport of sodium and chloride to the shoot. Journal of Experimental Botany. 2005;**56**:703-712

[81] Martínez-Ballesta MC, Alcaraz-López C, Muries B, Mota-CadenasC, CarvajalM. Physiological aspects of rootstock-scion interactions. Scientia Horticulturae Amsterdam. 2010;**127**:112-118

[82] Prior N. North Dakota State University. Grad Study Criminology and Criminal Justice. 2020;**5**:212-213

[83] Lee JM, Kubota C, Tsao SJ, Bie Z, Echevarria PH, Morra L, et al. Current status of vegetable grafting: Diffusion, grafting techniques, automation. Scientia Horticulturae [Internet]. Amsterdam: Elsevier B.V.; 2010;**127**:93-105. DOI: 10.1016/j.scienta.2010.08.003

[84] Martinez-Rodriguez MM, Estañ MT, Moyano E, Garcia-Abellan JO, Flores FB, Campos JF, et al. The effectiveness of grafting to improve salt tolerance in tomato when an "excluder" genotype is used as scion. Environmental and Experimental Botany. 2008;**63**:392-401

[85] Zrig A, AbdElgawad H, Touneckti T, Ben Mohamed H, Hamouda F, Khemira H. Potassium and calcium improve salt tolerance of *Thymus vulgaris* by activating the antioxidant systems. Scientia Horticulturae. Amsterdam. 2021;**277**

[86] Tounekti T, Vadel AM, Oñate M, Khemira H, Munné-Bosch S.

Salt-induced oxidative stress in rosemary plants: Damage or protection? Environmental and Experimental Botany. 2011;**71**:298-305

[87] Massai R, Remorini D, Tattini M. Gas exchange, water relations and osmotic adjustment in two scion/rootstock combinations of prunus under various salinity concentrations. Plant and Soil. 2004;**259**:153-162

[88] Fullana-Pericàs M, Conesa M, Pérez-Alfocea F, Galmés J. The influence of grafting on crops' photosynthetic performance. Plant Science[Internet]. Elsevier; 2019:110250. DOI: 10.1016/j.plantsci.2019.110250

[89] Pérez-Alfocea F, Albacete A, Ghanem ME, Dodd IC. Hormonal regulation of sourcesink relations to maintain crop productivity under salinity: A case study of root-to-shoot signalling in tomato. Functional Plant Biology. 2010;**37**:592-603

[90] Villalta I. Genetic analysis of Na^+ and K^+ concentrations in leaf and stem as physiological components of salt tolerance in Tomato Genetic analysis of Na^+ and K^+ concentrations in leaf and stem. 2008

[91] Rajendran K, Tester M, Roy SJ. Quantifying the three main components of salinity tolerance in cereals. Plant, Cell & Environment. 2009;**23**:237-249. DOI: 10.1111/j.1365-3040.2008.01916.x

[92] Ziska LH, DeJong TM, Hoffman GF, Mead RM. Sodium and chloride distribution in salt-stressed Prunus salicina, a deciduous tree species. Tree Physiology. 1991;**8**:47-57

[93] Mauro RP, Pérez-Alfocea F, Cookson SJ, Ollat N, Vitale A. Editorial: Physiological and molecular aspects of plant rootstock-scion interactions. Frontiers in Plant Science. 2022;**13**:1-5

[94] Lu X, Liu W, Wang T, Zhang J, Li X, Zhang W. Systemic long-distance signaling and communication between rootstock and scion in grafted vegetables. Frontiers in Plant Science. 2020;**11**:1-11

[95] Stahl-Biskup E, Saez F. Thyme the genus thymus. Med Aromat Plants–Industrial Profiles, 24. 2002;**15**(xii):330

[96] Zrig A, Ferreira JFS, Serrano M, Valero D, Tounekti T, Khemira H. Polyamines and other secondary metabolites of green-leaf and red-leaf almond rootstocks trigger in response to salinity. Pakistan Journal of Botany. 2018;**50**:245-250

[97] Ofosu M, Emmanuel BG, Khalilia WM, Yahaya MS, Salisi MS, Md. Isa NM, et al. Recent Research Advances in Biology. 2020;**4**

Chapter 7

Perspective Chapter: An Insight into Abiotic Stresses in Pigeonpea – Effects and Tolerance

Megha and Nisha Singh

Abstract

Cajanus cajan (L.) Millsp. is an adaptable, resilient, and nutrient-dense grain legume with qualities that can help agricultural systems become more sustainable in subtropical and tropical regions. Extremes in moisture, temperature, photoperiod, and mineral-related stressors are the most significant that encounter throughout the life cycle. Different stress slows down the plants' growth by upsetting its typical physiology and morphology. Inefficient conditions can occur sequentially or simultaneously in environments, and plants have evolved defense mechanisms to continue to grow and survive under less-than-ideal edaphic and climatic factors. Although various genotypes of pigeonpea have been identified that are tolerant to heat, drought, and waterlogging, new empirical evidence reveals that genotypic changes have been detected for all of the abiotic stimuli in this crop. Furthermore, to enhance tolerance, breeding techniques or methods such as marker features, including extensive hybridization, double haploids, tissue culture, somaclonal variants, genetic transformation, and marker-assisted breeding, have been employed to lessen the effect of these stressor. These methods help in the development of enhanced germplasm with abiotic stress tolerance and disease resistance, resulting in higher crop quality and production. This chapter focuses on different abiotic stressors and the methods that have been employed to help pigeonpea to overcome environmental constraints.

Keywords: abiotic stress, climatic changes, drought, genetic approach, molecular marker selection

1. Introduction

Legumes are well known for their nutritional and health benefits, as well as their contribution to agricultural system sustainability. They are the most important single source of vegetable protein in human diets and cattle feed (forages) [1]. Legumes are frequently used as an intercrop (e.g., paired with cereals) or in crop rotation in farming systems, resulting in a reduction in pests, diseases, and weed populations while increasing overall farm production and income for smallholder farmers. Other than

Cajanus cajan		
	Genomics resources	**References**
Species	Diploid	[3]
Genome size	833.07 Mbp	[4]
Genetic maps	Reference genetic map, six intraspecific maps, one consensus map	[5]
	DArT based maternal and paternal maps	[6]
SNPs array	50 K Affymetrix Axiom	[7]
	Genotyping-by-sequencing (GBS)	[6]
	Restriction site-associated DNA sequencing (RAD)	[8]
Database	PpTFDB, Pipemicrodb	[9]
Number of genes and ESTs	48,680 and 25,640	[3]
Whole genome sequencing	Reference genome sequence	[3, 10]
	WGRS	[11, 12]
Genetic purity testing kits	SSR assay	[13, 14]

Table 1.
List of genetic resources of pigeonpea.

commercial and economic importance, legumes have gotten less attention than cereals in terms of increasing agricultural production. A variety of abiotic stresses are threatening the legume crops [2]. Studies on stress tolerance processes have led to the identification of characters related with tolerance in plants, as well as the molecular regulation of stress-responsive genes. Some of these researches have paved the way for new opportunities to investigate the molecular basis of plant stress responses and find novel features and associated genes for agricultural plant genetic improvement (**Table 1**) [15].

Cajanus cajan also named as Pigeonpea, arhar, tur, red gram, is a major pulse crop of the world's semi-arid regions and India's second most important pulse crop after chickpea. It is high in protein (21–28%), carbohydrates, vitamins, fats, and minerals [4]. Pigeonpea has become an important crop in India throughout time, with attempts being undertaken to produce high yielding varieties by conventional breeding and biotechnology approaches [16]. Plants have evolved complex signaling pathways that include receptors, secondary messengers, phytohormones, and signal transducers to detect different stresses and adapt to changing environmental conditions. These inherent processes promote stress signal transduction and the activation of stress-responsive gene expression in order to maintain plant growth and productivity [17].

Pigeonpea breeding has been more difficult than in other dietary legumes due to crop specific features and a very sensitive nature [18]. For more than five decades, low productivity and lack of stability have been major production challenges in this crop. This scenario is caused by abiotic stressors, in addition to genetic and agronomic factors. This dilemma can now be turned around by simultaneously reducing crop losses and increasing crop yielding capabilities [19]. This hardy crop is subjected to a variety of abiotic stresses, including moisture (waterlogging/drought), temperature, photoperiod, and mineral (salinity/acidity) stress (**Table 2**) [25]. Drought and heat stress, two important abiotic stress elements affecting crop loss and yield, are notable effects of climate change. Drought disrupts the pigeonpeas' symbiotic association, reducing growth and finally leading to lower crop production [26]. The tension exerted on the

Abiotic stress	Genotypes	Tolerance mechanism	Reference
Waterlogging	ICPL 84023, Asha	Lenticels development, more root biomass and adventitious root.	[20]
Drought	LRG 30, ICPL 85063, ICPL 332	High RWC, pods/plant and HI.	[21]
Low Temperature	IPA 7-2, Bahar, and MAL 19	Ability to flower and pod setting under low temperature.	[22]
Salinity	C11, ICPL 227, WRP1, GS1 and TS3, UPAS 120 and ICPL 151	Reduced translocation of Na and Cl from root to shoot.	[3]
Aluminum toxicity	IPA 7-10 and T 7, 67 B and GT 101E	Aluminium exclusion.	[23]
Cold	ICPL 87119	Involved in seed germination and metabolism	[24]

Table 2.
Different pigeonpea genotypes tolerant to abiotic stress and their mechanism [18].

northern and north-eastern areas of India where temperature extremes (too low/too high) during the reproductive stage affect the production rate [27].

Other than temperature and drought, aluminum toxicity in acidic soil is also a constraint for production. Some regions of Haryana and Punjab, where the pigeonpea is affected by waterlogging, soil erosion and salinity pressures [28]. All these factors have a significant impact on productivity, yet few changes have been made in genotypes which are resistant to these abiotic stresses. Hence, the purpose of the present study is to examine the available information on abiotic stresses and discuss approaches to improve pigeonpea resistance to these constraints.

2. Major stresses influencing *C. cajan* productivity

2.1 Waterlogging

Waterlogging is a major abiotic stress affecting pigeonpea production where annual rainfall is 600–1500 mm [29]. The primary biological consequence of waterlogging is a lack (hypoxia) or full lack (anoxia) of oxygen in the soil, which creates anaerobic conditions and limits plant growth and development, thus seed yield. Oxygen shortage causes electrolyte leakage, exposing the cell to the outside environment, which can lead to lipid and nucleic acid peroxidation and, eventually, death [30]. Previous studies found that proline accumulation, starch content, and effective H_2O_2 detoxification are among the significant biochemical alterations that play a major role in waterlogging resistance in pigeonpea genotypes. During monsoon, this plant is susceptible to phytophthora blight disease due to waterlogging and hence causes yield losses [31].

Another study identified that high nitrogen uptake and development of aerenchyma in ICPL-84023 enabled it to sustain growth under waterlogging [32]. Total reducing sugars, superoxide dismutase, membrane stability index, number of pods per plant, pod dry weight, and seed yield, are some biochemical changes that can be affected by waterlogging [33]. Crossings between tolerant and sensitive lines revealed greater genetic variety than crosses between tolerant lines, implying the possibility of genetic improvement for this crop. Lines derived from crosses involving *C. acutifolius* have the unique property of enhancing tolerance under water logging circumstances for resource poor farming communities [29]. It is estimated that agriculture would be impacted globally by these forecasted climate changes. A better understanding of this legume crop resource and their characterization in terms of desirable traits for climate change adaptation are essential for the use of adapted *C. cajan* genetic resources in strengthening the resilience of future production systems [34].

2.2 Drought

Pigeonpea is a rainfed crop grown during the kharif season. Because of its deep root structure, it is considered a drought-tolerant legume [25]. It can suffer from early and terminal drought stress due to its deep and broad root structure [26]. The germplasm has a wide range of Osmotic adjustment variation (0.2–1.6 MPa), with some accessions reaching 5.0 MPa. Some varieties, such as Bahar, BSMR 853, and ICP 84031, have demonstrated increased osmotic adjustment under drought conditions [35]. In response to drought conditions, relative water content (RWC) of leaves and dehydration tolerance are crucial (**Figure 1**). Drought resistant breeding should be done under true moisture-deficit conditions using agronomic traits such as pods per plant, seeds per pod, seed size, and seed production per plant. Therefore, physiological interactions, as well as high mean seed yield, should be employed to identify superior genotypes for low-soil-moisture situations [23, 36].

To understand the molecular mechanism for drought response in pigeonpea, a study has been performed on ICP151, ICPL8755 and ICPL227, where 51 genes were selected using Hidden Markov Model (HMM) to identify protein domain responsible for stress-responsive genes. Ten genes of U-box proteins, H^+ antiporter proteins, and universal stress proteins were studied out of 51 drought genes (*AuspA*). These genes offer the way for molecular research into drought resistance [37]. The identified genes can also be validated at the sequence level in various genetic backgrounds to identify

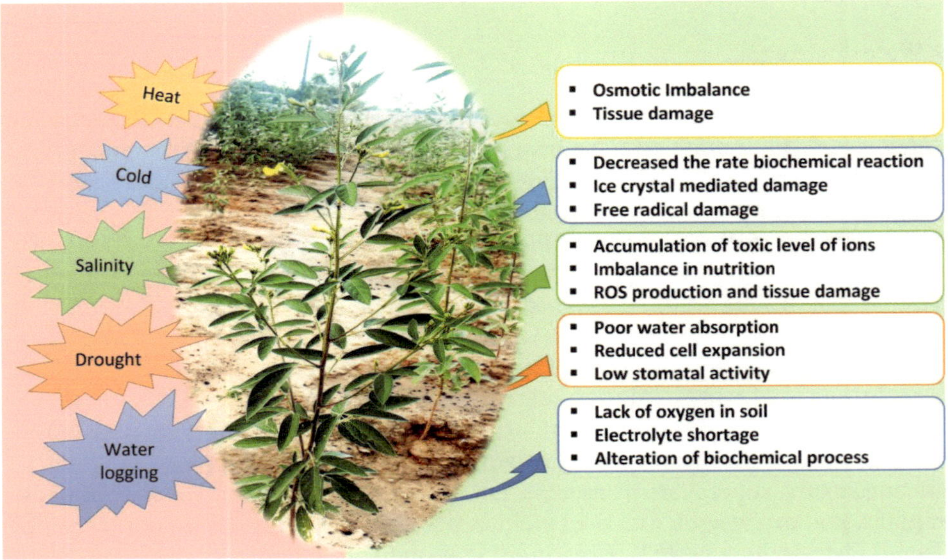

Figure 1.
Abiotic stresses in pigeonpea with their effects.

the presence of sequence variations for the formation of gene-based markers for crop improvement and the development of breeding lines and hybrids that are more tolerant through genomics-assisted breeding [38, 39]. As drought stress stagnates the food security over the globe, it is important to develop new varieties to achieve a proper amount of yield with maintained quality under such climatic perturbation. Strategies should develop where pigeonpea could enhance physicochemical capability of their cells to continue metabolism at low leaf water status [40].

2.3 Soil salinity

Salt stress is a significant constraint to the productivity of the nutritional rich pigeonpea. India accounts for more than 85% of global production and consumption of this legume crop. Excess Na^+ accumulation during salt stress interferes with hydrogen bonding and polar interactions, causing protein and nucleic acid structure to be disrupted. Thus, the total soluble protein content of stressed pigeonpea plants was found to be significantly lower [24]. The moisture content and succulence of *C. cajan* decreased dramatically as salinity increased, indicating a loss of turgor. When subjected to increasing salinity, this crop reduces water content in order to reach low osmotic potential. Salinity was performed to extend the 50% flowering stage by 1–2 weeks while also delaying the peak flowering stage. It increases floral shedding, lowering the effective quantity and weight of pods, and lastly lowering the number of seeds, lowering production [41].

Previous studies observed that the salt tolerance gene, *CcCYP*, is responsible for upregulated salt tolerance in root, whereas *CcCDR* was upregulated in shoot [42]. To make this legume crop resilient to salt stress, a better understanding of the molecular networks, in particular epigenetic regulation of gene expression, would be beneficial [23]. The potential of producing salt-tolerant lines of pigeon pea through genetic engineering has not been thoroughly studied. There is only one occurrence where transgenic pigeon pea plants were given salt tolerance through overexpression of the mutant 1-pyrroline-5-carboxylate synthetase gene (P5CSF129A) from *Vigna aconitifolia* [43]. These lines are notably salt tolerant. The identification of novel molecular targets that can be exploited by transgenic technologies would undoubtedly benefit from genome-wide association studies (GWAS) that uncover gene expression profiles in salt-stressed pigeon pea. It is also possible to use genomics-assisted intensive breeding to find quantitative trait loci and potential markers in salt-tolerant pigeon pea cultivars [44].

2.4 Metal toxicity

Changes in the environment are most likely to have a significant impact on how plants evolve, mostly through interfering with the process through mutations, gene flow, and evolution. Heavy metals are the major environmental changes/pollutants and their toxicity is a growing concern for ecological, evolutionary, nutritional and environmental reasons. These contaminants have a negative impact on the environment, impair agricultural output, and pose serious health risks to living organisms [45]. Metals exerts several effects on legume crop generated by elements such as Cd, Cu, Al, Hg, Pb and As, among others.

Cadmium (Cd), the most dangerous heavy metals because of their great mobility, non-degradability, and toxicity to plants as well as animals [46]. Excessive Cd^{2+} accumulation in plants can result in severe phytotoxicity as well as a variety of physiological, morphological, and biochemical toxic effects on plant attributes such as pigment destruction, photosynthetic and respiration process inhibition, reduced nutrient uptake, overproduction of reactive oxygen species (ROS), enzyme and gene suppression, growth inhibition, and even plant death [47, 48].

Copper (Cu) is a vital element for plants since it helps with several physiological processes such as mitochondrial respiration, photosynthetic electron transport, and cell wall metabolism [49]. However, due to its redox characteristics, it is harmful to plants in large quantities (**Figure 2**). Excessive amounts impede plant growth, interfere with photosynthetic and respiratory activities, reduce nutrient uptake, target the membrane transport system, and produce excessive amounts of ROS [50]. Copper concentrations in the soil gradually lowered pigeonpea secondary metabolite biosynthesis (phenolics and flavonoids). Under Cu stress, pigeonpea had severe oxidative damage, as evidenced by higher levels of MDA (Malondialdehyde contents), hydrogen peroxide, and electrolyte leakage. Antioxidant enzymes (Superoxide dismutase, Peroxidase dismutase, Catalase and Glutathione peroxidase) and proline content were considerably increased with increasing Cu concentration to reduce oxidative damage [51].

Mercury contamination has emerged as a critical modern environmental issue. Its treatment highly reduced seed germination. Mercury chloride was found to be very harmful to seedling growth of legume crops. Plants grown at various levels of cadmium revealed a considerable drop in the length of shoots and roots, yellowing and ultrastructural abnormalities of the leaves, and a significant decrease in the essential oil content [52]. This metal exists in both organic and inorganic forms, and both are extremely dangerous. Its concentration in soil and water is an issue due to the widespread use of mercury-containing chemicals, fungicides, algaecide, paper pulp

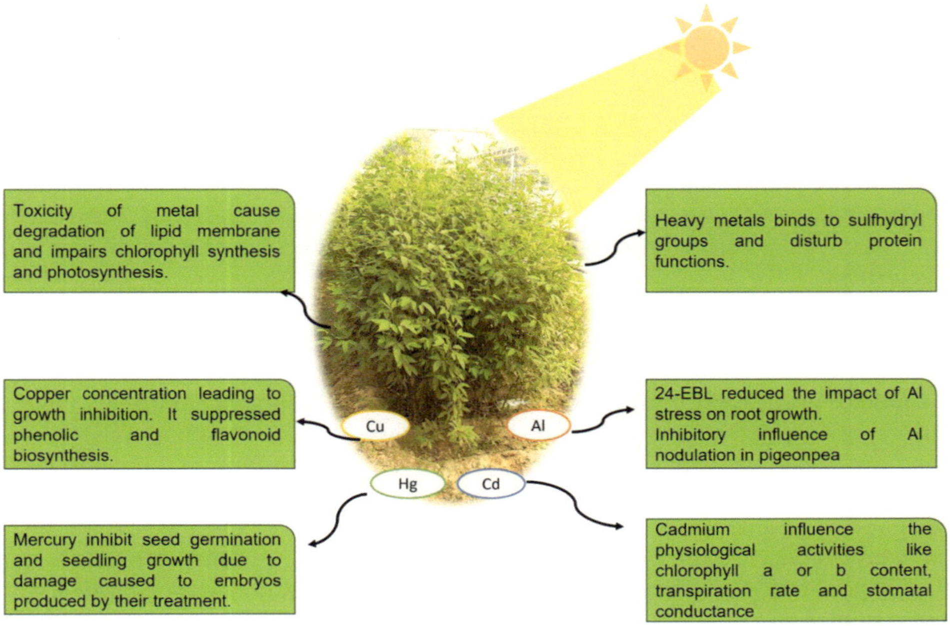

Figure 2.
Toxic effect of different heavy metals on pigeonpea.

industries, and agriculture. Mercury released into the near environment may penetrate pigeonpea and other crops that humans eat, affecting human health. Therefore, it is critical to reduce the use of mercury in industries, as well as mercury-containing insecticides and fungicides [53].

Aluminum is the third most prevalent element in the earth's crust (after oxygen and silicon). The presence of poisonous Al^{3+} cations in acidic soils (pH 5.0) is a major constraint to agricultural productivity worldwide. The excess of Al is a major soil limitation to food and biomass production [54]. The suppression of root extension is the first sign of Al toxicity, which has been postulated to be produced by a variety of mechanisms, including Al interactions with the plasma membrane or the symplast. Aluminum poisoning has a negative impact on root growth and interferes with water and mineral nutrient intake [54, 55]. Pigeonpea plants cultivated in Al-challenged soil have lower nodulation. However, the use of 24-EBL inhibited the effect of Al on nodulation. Rhizobium multiplication and nodule development were reported to be more sensitive aspects of the symbiotic interaction to excess Al. Al poisoning caused a significant decrease in chlorophyll concentration. The use of 24-EBL on *C. cajan* plants significantly boosted photosynthetic pigments and counteracted the negative effects of Al^{+3} stress [18, 23, 56]. Plants have evolved various strategies to minimize metal-induced damage, including metal exclusion, compartmentalization, chelation, and a wide range of ROS-scavenging mechanisms, including antioxidant enzymes such as superoxide dismutase (SOD), catalase (CAT), peroxidase (POX), glutathione reductase (GR), ascorbate peroxidase (APX), as well as non-enzymatic antioxidants such as phenol [18, 51].

2.5 Temperature

2.5.1 Cold stress

During the winter season in northern India, pigeonpea suffers from low temperature stress (December–January). If the minimum temperature goes below 5°C, stress impacts plant growth, survival, and reproductive capacity [57]. At freezing temperatures, intracellular water condenses into ice, causing cell contraction within the plant, resulting in wilting and plant death [18, 58]. Initial research at IIPR Kanpur also revealed genotypic differences in cold tolerance in pigeonpea. Because knowledge on cold stress and its impact on the pigeonpea crop is scarce, screening a large number of pigeonpea genotypes for low temperature tolerance under controlled temperature conditions is still required to confirm and generate precise genetic information [59].

2.5.2 Heat stress (HS)

Pigeonpea is a grain legume that is resistant to climate change. Though the ideal temperature for produced is 25–35°C, wild relatives grow at temperatures ranging from 18 to 45°C [60]. Heat Stress (HS) is the most serious abiotic threat to all legume crops. It reduces plant biomass build-up, resulting in lower yield, particularly in tropical and subtropical environments [61]. A prior analysis stated that a 1°C increase in maximum temperature during crop season could result in a 20.8% decrease in pigeonpea output. HS causes critical protein complexes to dissociate and the production of Reactive Oxygen Species (ROS) [62]. Plants tend to up-regulate the genes encoding molecular chaperones and signaling molecules in response to HS, thereby regulating a chain of events that lead to HS responses [63].

3. Mitigate the climatic change for *C. cajan* production

Food production, security and climate change are all interconnected and hence affecting living systems. Long-term changes cause the entire weather pattern to alter, and also increase temperature, unpredictable rainfall, floods and a rise in sea level. India and other developing nations struggle to produce enough food to feed their expanding populations [23, 63]. Legumes, especially pulses make up the majority of the food on an Indian meal platter. Over a few decades, Pigeonpea in India has transitioned from being an orphan crop to a cash crop. Its production as a main crop is constrained by its lengthy maturation period and low yield [64]. This crop can withstand prolonged periods of drought and are well adapted to rain-fed conditions. They require little soil moisture to maintain themselves and generate a respectable amount of yield. However, this legume crop is sensitive to high temperatures and waterlogging. The effects of shifting climatic conditions on arhar that are rainfed are significant [64, 65]. According to reports, pulses are especially susceptible to heat stress during the bloom stage; just a few days of exposure to high temperatures ((30–35°C) can result in significant yield losses due to flower drop or pod damage. The crop's ability to grow in a larger range of latitudes and altitudes has been constrained by shifting rainfall patterns, rising yearly temperatures, and irregular climatic trends. However, there is no denying that the crop has the potential to support food security, nutrition, forage, and income production [66, 67].

Indian farmers have long waited for early-maturing pigeonpea cultivars that are compatible with their farming practices and produce higher yields with little inputs. The super-early varieties (ICPL 11255, ICPL 20340, and ICPL 20338) that ICRISAT's pigeonpea breeding team recently created are luring farmers from numerous states, including Maharashtra, Odisha, Karnataka, Telangana, and Andhra Pradesh [https://www.icrisat.org/]. Given their photo- and thermo-insensitivity and capacity to grow in a larger range of latitudes (30°N) and altitude (1250 msl), such as in Uttarakhand, Rajasthan, Odisha, and Punjab, these cultivars have the potential to flourish in varied agro-ecologies. Creating short-lived variants has an added benefit. They may be cultivated with minimal inputs post-rainy season or off-season, giving farmers in dryland areas of India an extra source of income [68].

Gene mining for abiotic stress tolerance, restructuring plant types for climate-vulnerable regions, changing cropping patterns, effective nutrient and water management, seed banks for alternative legume crops, watershed management, and micro-irrigation facilities are some of the better options to address climate change-related issues [42, 67, 69]. Furthermore, crop improvement strategies could be enhanced to mitigate climate changes by developing climate resilient varieties, reducing crop duration, adopting diversification in practices, improving crop specific practices, reducing greenhouse gas emission and use of biofertilizers. Therefore, more effective agronomic techniques have a huge potential to counteract the negative effects of climate change on arhar production. Adopting suggested management measures helps agriculture not only conserve soil and water, but also increases soil organic carbon levels and lessens the effects of climate change [70].

4. Genetic enhancement in abiotic stress tolerance in pigeonpea

Genomics is concerned with the physical integrity of the genome, with the purpose of identifying, diagnosing, and regulating genetic traits throughout the chromosomes. We are now considering certain genetic advances to better understand abiotic stress tolerance in Pigeonpea. Specific trait markers for blooming, fertility, and resistance to sterility mosaic disease. QTL mapping, association mapping for candidate genes, transcriptome assembly, and genome sequencing technologies can be used to identify yield factors [71].

5. Quantitative trait loci mapping

Abiotic stress resistance heredity is a complicated process, QTL mapping, genetic and linkage mapping of genomic regions relevant to tolerance, is the most preferred way of discovering QTL. QTL analysis allows researchers to investigate the genetic structure of a trait. QTLs can discover genomic regions associated with the expression of the characteristic under investigation [72]. Different types of bi-parental populations are used for QTL mapping and the discovery of marker-trait associations. These populations include recombinant inbred lines (RILs), near isogenic lines (NILs), doubled haploids (DHs), multiparent advanced generation inter-cross (MAGIC), nested association mapping (NAM), and association mapping (AM) on wider panels [73]. Chickpea, pigeonpea, lentil (*Lens culinaris*), and groundnut have successfully used advanced backcross quantitative trait loci (AB-QTL). The resolution for locating novel genes, alleles, and QTLs is improved when this bi-parental mapping population is combined with GBS and GWAS [39, 74]. PEG/water deprivation stress conditions were used to create a collection of ESTs from entire plant tissues of pigeonpea [75]. From pigeonpea plants treated with 10% polyethylene glycol, two subtracted cDNA libraries were created (PEG-6000). Among the many ESTs found, three stress-responsive genes, *CcHyPRP* [76], *CcCDR* [77], and *CcCYP* [75], demonstrated extraordinary resistance to different abiotic stimuli in transgenic Arabidopsis.

6. Molecular marker resources

With the advent of genomic research, new opportunities for genetic enhancement of complex traits like salinity and drought endurance have emerged. In comparison to traditional breeding, a combination of genomic approaches and molecular marker resources can significantly speed up the identification of individual-specific genes in breeding populations [78]. This explains the evolution of genetic marker technology from gel or hybridization approaches (DArT, SFP's) to sequence-based SSR and SNP markers. Diversity arrays technology (DArT), a hybridization based highly parallel genotyping protocol, has generated thousands of polymorphic loci in pigeonpea that were used for genetic diversity analysis and linkage mapping [6]. SNPs helps in the identification of haplotypes, and blocking such haplotypes would act as markers for the identification of relevant attributes utilizing allele mining approaches [79]. The 50 K Rice SNP50 array was developed for Illumina Infinium platform and has thousands of genome-wide SNPs with genic regions responsible for different genic regions. Further, this array was successfully used for variety verification and trait introgression. The 50 K Rice SNP50 chip plays an important role in both functional and genomics studies and molecular breeding [80, 81]. Similar analysis has been performed using a 62 K SNP array in pigeonpea germplasm. Incorporation of 746 disease resistance and defense response genes in the array with average 10 SNPs per gene will be useful for pathologists and breeders in identifying genes for abiotic stress resistance in pigeonpea [82, 83].

Utilizing diverse genomics resources and enhanced genotyping platforms, molecular breeding techniques like MAS (marker-assisted selection), MABC (marker-assisted backcross breeding), GS, and multivariate adaptive regression splines (MARS) allow for the effective use of legume crop genetic resources that contain important alleles and genes. For instance, four molecular markers (ICCM0249, TAA170, GA24, and STMS11) have been transmitted by MABC for the creation of chickpea types that can tolerate drought [84–87]. Such markers have been developed by understanding the genome-wide sequence variations and are effectively utilized for allele mining, characterizing germplasm for genetic improvement and genetic mapping of important agronomics traits.

7. Genome-wide association studies (GWAS)

The identification of candidate gene(s)/QTLs for complex characteristics is significantly assisted by GWAS. GWAS methods have been used to find small and minor genetic changes linked to a variety of biotic and abiotic stresses as well as crop agronomic traits [88–90]. GWAS analyzes the entire genome for QTLs and requires for genome-wide markers. Through the GWAS method, different QTLs were also discovered for several abiotic stress-tolerant genes. The genetic resources and gene(s)/QTLs for morphological, quality, and biotic and abiotic stressors have recently been enriched in pigeonpea [89, 90]. Through MAS, the yield traits as well as the detected QTLs/gene(s), such as pod borer and Phytophthora stem blight resistance genes, have been successfully introgressed into the cultivated varieties of pigeonpea [91]. To speed up genetic gain, two high-density Affymetrix Axiom genotyping chips have recently been created. 103 lines were studied using a 56 K *Cajanus* SNP chip to examine genetic diversity. The SNPs lack haplotype information and are distributed at random [40].

A 62 K genic-SNP chip called "CcSNPnks" has recently been created using the resequencing of 45 different genotypes. Additionally, the 'CcSNPnks' chip array will be helpful for gene-based association studies and high-resolution mapping of yield-related QTLs. With the use of these high throughput genotyping arrays, many samples may be genotyped quickly, and the analysis of the primary genotyping data is also relatively simple [92]. In pigeonpea from diverse sets of wild and cultivated genetic backgrounds, this led to the discovery of the most effective genomic loci (genes) associated with abiotic and biotic stress related genes [4, 83].

8. Transcriptomics profiling

Transcriptomic tools scan, provide gene-expression and protein expression levels in real time, making them important in plant improvement in this advanced era. The development of next-generation sequencing technology has made it possible and reliable to sequence plant species [93]. Furthermore, transcriptomics technologies help to understand gene and protein levels. According to the findings of several research, not all genes are turned on or off at the same time; hence, the metabolism adopts a complex phenotype that cannot be determined by genotype [38, 94]. As of December

26th, 2014, 25,577 ESTs for pigeonpea were discovered at NCBI (National Centre for Biotechnology Information). CcTAv1 transcriptome assembly contigs were created with 1, 27, 754 TUS (Tentative Unique Sequences) and were then upgraded with Illumina GAIIX by 454 platforms to construct CcTav2 transcriptome assembly contigs with four data groups and 21, 434 transcriptome assembly contigs (TAC's) [95–97]. The expression of WRKY genes in two different genotypes was examined in leaf and root tissue in response to drought and salt stress [98].

Furthermore, Comparative transcriptome analysis and biochemical tests showed that *Cajanus* species' responses to heat stress varied widely. The most thermotolerant of the examined species was *C. scarabaeoides*, followed by *C. cajanifolius*, *C. cajan*, and *C. acutifolius*. When under heat stress, a significant number of genes have been studied that undergo alternative splicing in a species-specific pattern. Chlorophyll content, electrolyte leakage assay, histochemical assay, and gene expression profiling analysis all demonstrated that *C. scarabaeoides* possesses adaptive traits for heat stress tolerance [61]. It would help breeders find promising candidate genes and appropriate features for creating and boosting legume crop productivity under abiotic challenges [24]. In depth analysis of the transcriptomics would be definitely fascinating for better perception of pigeonpea.

9. Conclusion and future prospects

Food production will face severe hurdles in the near future due to a gradual drop in soil water and an increase in temperature. Drought and high temperature tolerant crops, such as pigeonpea, may be a viable option for ensuring food security. Efforts should be made to define the genetic resources of pigeonpea at both the phenotypic and molecular levels in order to uncover genetic variations that can be leveraged to generate improved cultivars. To achieve a consistent rise in pigeonpea productivity, existing breeding efficiency must be improved.

In order to focus on trait associated marker study, new methodologies such as transcriptome assembly, MAGIC, and NAM populations were developed. It is feasible to introduce genes from wild species to commercially farmed types using cutting-edge advanced backcross-QTL techniques. In the future, efforts should be made to concentrate on phenotypic approaches that are affordable, high throughput, and effective. Innovative breeding designs that are supported by relevant genomic technology will be critical in modernizing breeding programmes. Current genetic advances in pigeonpea for resistance to abiotic stress will also considerably benefit hybrid breeding. Furthermore, intense attempts are being made using in vitro techniques to find complicated abiotic stress features, foreign gene introgression facilitated by embryo rescue, and quick fixation of stress tolerant recombinants via doubled haploid breeding. These procedures, together with more efficient screening methods, demand special attention in the coming days to make pigeonpea farming an attractive, profitable, and feasible option for the world's pulse farmers.

Acknowledgements

NS acknowledges the Department of Science and Technology; Government of India for the DST INSPIRE Faculty Award (DST/INSPIRE/04/2018/003674).

Author contributions

NS: conceived the study, edit the manuscript. M: contributed to the writing and editing of the manuscript. Both authors contributed to the writing, editing, and approved the manuscript.

Conflicts of interest

We have no conflicts of interest to disclose.

Author details

Megha and Nisha Singh*
Gujarat Biotechnology University (GBU), Gandhinagar, Gujarat, India

*Address all correspondence to: singh.nisha88@gmail.com; nisha.singh@gbu.edu.in

References

[1] Anwar A, Sinha A, Sharif A, Siddique M, Irshad S, Anwar W, et al. The nexus between urbanization, renewable energy consumption, financial development, and CO_2 emissions: Evidence from selected Asian countries. Environment, Development and Sustainability. 2022;**24**:6556-6576

[2] Pandey R, Aretano R, Gupta AK, Meena D, Kumar B, Alatalo JM. Agroecology as a climate change adaptation strategy for smallholders of Tehri-Garhwal in the Indian Himalayan region. Small-scale Forestry. 2017;**16**: 53-63. DOI: 10.1007/s11842-016-9342-1

[3] Kumar P, Choudhary M, Halder T, Prakash NR, Singh V, Sheoran S. Salinity stress tolerance and omics approaches: Revisiting the progress and achievements in major cereal crops. Heredity. 2022:1-22

[4] Singh N, Rai V, Singh NK. Multi-omics strategies and prospects to enhance seed quality and nutritional traits in pigeonpea. The Nucleus. 2020;**63**(3):249-256

[5] Varshney RK, Kudapa H, Roorkiwal M, Mahendar T, Pandey MK, Saxena RK, et al. Advances in genetics and molecular breeding of three legume crops of semi-arid tropics using nextgeneration sequencing and high-throughput genotyping technologies. Journal of Biosciences. 2012;**37**:811-820

[6] Yang SY, Saxena RK, Kulwal PL, Ash GJ, Dubey A, Harper JD, et al. The first genetic map of pigeon pea based on diversity arrays technology (DArT) markers. Journal of Genetics. 2011;**90**(1):103-109

[7] Saxena RK, Kale SM, Kumar V, Parupalli S, Joshi S, Singh VK, et al. Genotyping-by-sequencing of three mapping populations for identification of candidate genomic regions for resistance to sterility mosaic disease in pigeonpea. Scientific Reports. 2017;**7**:1813

[8] Saxena RK, Rathore A, Bohra A, Yadav P, Das RR, Khan AW, et al. Development and application of high-density Axiom-*Cajanus* SNP array with 56 K SNPs to understand the genome architecture of released cultivars and founder genotypes. The Plant Genome. 2018;**11**:180005

[9] Arora S, Mahato AK, Singh S, Mandal P, Bhutani S, Dutta S, et al. A high density intraspecifc SNP linkage map of pigeonpea (*Cajanas cajan* L. Millsp.). PLoS One. 2017;**12**:e0179747

[10] Sarika VA, Iquebal MA, Rai A, Kumar D. PIPEMicroDB: Microsatellite database and primer generation tool for pigeonpea genome. Database. 2013;**2013**

[11] Singh NK, Gupta DK, Jayaswal PK, Mahato AK, Dutta S, Singh S, et al. The first draft of the pigeonpea genome sequence. Journal of Plant Biochemistry and Biotechnology. 2011;**21**:98-112

[12] Kumar V, Khan AW, Saxena RK, Garg V, Varshney RK. First generation hapmap in *Cajanus* spp. reveals untapped variations in parental lines of mapping populations. Plant Biotechnology Journal. 2016;**14**:1673-1681

[13] Singh VK, Khan AW, Saxena RK, Kumar V, Kale SM, Sinha P, et al. Next-generation sequencing for identification of candidate genes for Fusarium wilt and sterility mosaic disease in pigeonpea (*Cajanus cajan*). Plant Biotechnology Journal. 2016;**14**:1183-1194

[14] Saxena RK, Saxena K, Varshney RK. Application of SSR markers for molecular characterization of hybrid parents and purity assessment of ICPH 2438 hybrid of pigeonpea [*Cajanus cajan* (L.) Millspaugh]. Molecular Breeding. 2010;**26**:371-380

[15] Rane J, Singh AK, Kumar M, Boraiah KM, Meena KK, Pradhan A, et al. The adaptation and tolerance of major cereals and legumes to important abiotic stresses. International Journal of Molecular Sciences. 2021;**22**(23):12970

[16] Varshney RK, Saxena RK, Upadhyaya HD, Khan AW, Yu Y, Kim C, et al. Whole-genome resequencing of 292 pigeonpea accession identifies genomic regions associated with domestication and agronomic traits. Nature Genetics. 2017;**49**(7):1082-1088

[17] Yokotani N, Sato Y, Tanabe S, Chujo T, Shimizu T, Okada K. WRKY76 is a rice transcriptional repressor playing opposite roles in blast disease resistance and cold stress tolerance. Journal of Experimental Botany. 2013;**64**:5085-5097

[18] Choudhary AK, Sultana R, Pratap A, Nadarajan N, Jha UC. Breeding for abiotic stresses in pigeonpea. Journal of Food Legumes. 2011;**24**(3):165-174

[19] Ali A, Altaf MT, Nadeem MA, Karaköy T, Shah AN, Azeem H, et al. Recent advancement in OMICS approaches to enhance abiotic stress tolerance in legumes. Frontiers in Plant Science. 2022;**13**

[20] Sarode SB, Singh MN, Singh UP. Genetics of waterlogging tolerance in pigeonpea [*Cajanus cajan* (L.) Millsp.]. Indian Journal of Genetics and Plant Breeding. 2007;**67**:264-265

[21] Singh MN, Singh RS. Inheritance of pod setting under low temperature in pigeonpea. Indian Journal of Genetics and Plant Breeding. 2010;**70**:277-280

[22] Karajol K, Naik GR. "Seed germination rate" as a phenotypic marker for the selection of NaCl tolerant cultivars in pigeonpea (*Cajanus cajan* (L.) Millsp.). World Journal of Science and Technology. 2011;**1**:01-08

[23] Choudhary AK, Sultana R, Vales MI, Saxena KB, Kumar RR, Ratnakumar P. Integrated physiological and molecular approaches to improvement of abiotic stress tolerance in two pulse crops of the semi-arid tropics. The Crop Journal. 2018;**6**(2):99-114

[24] Singh A, Singh PK, Sharma AK, Singh NK, Sonah H, Deshmukh R, et al. Understanding the role of the WRKY gene family under stress conditions in pigeonpea (*Cajanus cajan* L.). Plants. 2019;**8**(7):214

[25] Settele J, Bishop J, Potts S. Climate change impacts on pollination. Nature Plants. 2016;**2**:16092

[26] Anjum SA, Ashraf U, Zohaib A, Tanveer M, Naeem M, Ali I. Growth and development responses of crop plants under drought stress: A review. Zemdirbyste-Agriculture. 2017;**104**:267-276

[27] Gupta S, Pandey S. Exploiting the potential of plant growth-promoting rhizobacteria in legume production. In: Abiotic Stress and Legumes. Elsevier; 2021. pp. 1-32

[28] Saxena KB, Hingane AJ, Choudhary AK, Bharathi H. A short-cut approach for breeding pigeonpea hybrids with tolerance to biotic and abiotic stresses. International Journal of Science and Research (IJSR). 2015;**4**(7):01-04

[29] Hingane AJ, Saxena KB, Patil SB, Sultana R, Srikanth S, Mallikarjuna N, et al. Mechanism of water-logging

tolerance in pigeonpea. Indian Journal of Genetics and Plant Breeding. 2015;**75**(2):208-214

[30] Duhan S, Sheokand S. Mechanism of waterlogging stress tolerance in pigeonpea plants: Biochemical and anatomical adaptation under waterlogging. In: Plant Ecophysiology and Adaptation under Climate Change: Mechanisms and Perspectives II. Singapore: Springer; 2020. pp. 89-105

[31] Singh Vinay Pratap, J. F. Srivasatava, and Ruchi Bansal. Biochemical responses as stress indicator to water logging in pigeon pea (*Cajanus cajan* L.). IJJB; 2017:300-305

[32] Bansal R, Srivastava JP. Effect of waterlogging on root anatomy and nitrogen distribution in pigeonpea (*Cajanus cajan* (L.) Millsp.). Indian Journal of Plant Physiology. 2017;**22**:130-134

[33] Yohan Y, Umamahesh V, Sudhakar P, Reddiramu Y, Sreedhar D. Influence of waterlogging on certain biochemical and yield parameters of pigeonpea (*Cajanus cajan* (L.) Millsp). International Journal of Pure & Applied Bioscience. 2017;**5**(4):1862-1868

[34] Zandalinas SI, Mittler R, Balfagón D, Arbona V, Gómez-Cadenas A. Plant adaptations to the combination of drought and high temperatures. Physiologia Plantarum. 2018;**162**:2-12

[35] Basu PS, Singh UMMED, Kumar ANIL, Praharaj CS, Shivran RK. Climate change and its mitigation strategies in pulses production. Indian Journal of Agronomy. 2016;**61**:S71-S82

[36] Kumar CVS, Mula MG, Singh IP, Saxena RK, Rao G, Varshney RK. Pigeon pea perspective in India. In: Proceedings of Philippine Congress. Phillipines; 2014. pp. 1-5

[37] Sinha P, Pazamala LT, Singh VK, Saxena RK, Krishamurty L, Azam S, et al. Identification and validation of selected universal stress protein domain containing drought-responsive genes in pigeon pea (*Cajanus cajan*). Frontiers in Plant Science. 2016;**6**:1-9

[38] Jha UC, Bohra A, Nayyar H. Advances in "omics" approaches to tackle drought stress in grain legumes. Plant Breeding. 2020;**139**:1-27

[39] Thudi M, Gaur PM, Krishnamurthy L, Mir RR, Kudapa H. Genomics-assisted breeding for drought tolerance in chickpea. Functional Plant Biology. 2014;**41**:1178-1190

[40] Saxena RK, Rathore A, Bohra A, Yadav P, Das RR, et al. Development and application of high-density Axiom Cajanus SNP Array with 56 K SNPs to understand the genome architecture of released cultivars and founder genotypes for redefining future pigeonpea breeding programs. The Plant Genome. 2018;**11**:180005

[41] Ahmed NAEEM, Ahmad RAFIQ. Salt stress responses of pigeon pea (*Cajanus cajan*) on growth, yield and some biochemical attributes. Pakistan Journal of Botany. 2016;**48**(4):1353-1360

[42] Awana M, Jain N, Samota MK, Rani K, Kumar A, Ray M, et al. Protein and gene integration analysis through proteome and transcriptome brings new insight into salt stress tolerance in pigeonpea (*Cajanus cajan* L.). International Journal of Biological Macromolecules. 2020;**164**:3589-3602

[43] Surekha CH, Kumari KN, Aruna LV, Suneetha G, Arundhati A, Kavi Kishor PB. Expression of the Vigna aconitifolia P5CSF129Agene in transgenic pigeonpea enhances proline accumulation and salt tolerance. Plant Cell, Tissue and Organ Culture. 2014;**116**:27-36

[44] Banerjee A, Ghosh P, Roychoudhury A. Salt stress responses in pigeon pea (Cajanus cajan L.). In: Pulse Improvement. Cham: Springer; 2018. pp. 99-108

[45] Masindi V, Muedi KL. Environmental contamination by heavy metals. Heavy Metals. 2018;**10**:115-132

[46] Zwolak A, Sarzyńska M, Szpyrka E, Stawarczyk K. Sources of soil pollution by heavy metals and their accumulation in vegetables: A review. Water, Air, and Soil Pollution. 2019;**230**:164

[47] Hu B, Shao S, Ni H, Fu Z, Hu L, Zhou Y, et al. Current status, spatial features, health risks, and potential driving factors of soil heavy metal pollution in China at province level. Environmental Pollution. 2020;**266**:114961

[48] Hayat K, Khan J, Khan A, Ullah S, Ali S, Fu Y. Ameliorative effects of exogenous proline on photosynthetic attributes, nutrients uptake, and oxidative stresses under cadmium in pigeon pea (*Cajanus cajan* L.). Plants. 2021;**10**:796

[49] Ullah S, Khan J, Hayat K, Abdelfattah EA, Salam U, Yu B, et al. Comparative study of growth, cadmium accumulation and tolerance of three chickpea (Cicer arietinum L.) cultivars. Plants. 2020;**9**:310

[50] Apori OS, Hanyabui E, Asiamah YJ. Remediation technology for copper contaminated soil: A review. Asian Soil Research Journal. 2018;**1**:1-7

[51] Hayat K, Khan A, Bibi F, Murad W, Fu Y, Batiha GES, et al. Effect of cadmium and copper exposure on growth, physio-chemicals and medicinal properties of *Cajanus cajan* L. (pigeon pea). Metabolites. 2021;**11**(11):769

[52] Patnaik A, Mohanty BK. Toxic effect of mercury and cadmium on germination and seedling growth of *Cajanus cajan* L. (pigeon pea). Annals of Biological Research. 2013;**4**(3):123-126

[53] Padmavathi S, Sujatha B. Assessment of mercury toxicity on seed germination, shoot and root growth of *Cajanus cajan* (L.) Millsp. International Journal of Creative Research Thoughts. 2021;**9**(1):428-436

[54] He H, He L, Minghua G. Role of microRNAs in aluminum stress in plants. Plant Cell Reports. 2014;**33**(6):831-836

[55] Thakur AK, Singh KJ. Seed emergence is more appropriate criterion for testing cadmium toxicity in soybean. International Journal of Food, Agriculture & Veterinary Sciences. 2014;**4**(2):129-134

[56] Sri ND et al. Amelioration of aluminium toxicity in pigeon pea [*Cajanus cajan* (L.) Millsp.] plant by 24-epibrassinolide. American. Journal of Plant Sciences. 2016;**7**(12):1618-1628

[57] Araujo SSS, Beebe M, Crespi B, Delbreil EM, Gonzalez V, Gruber I, et al. Abiotic stress response in legumes: Strategies used to cope with environmental challenges. Critical review. Plant Science. 2015;**34**:34237-34280

[58] NICRA Project. Kanpur: IIPR; 2016

[59] Alagupalamuthirsolai M, Vijayabharathi A. Genetic diversity analysis of pigeonpea [*Cajanus cajan* (L.) Millsp.] under cold stress conditions. Journal of Pharmacognosy and Phytochemistry. 2020;**9**(3):2048-2053

[60] Tiwari TN, Dutta D, Tewari K. Performance of pigeon pea [*Cajanus cajan* (L.) Millsp] genotypes under cold stress. Legume Research. 2022;**1**:1-6

[61] Ramakrishna G, Kaur P, Singh A, et al. Comparative transcriptome analyses revealed different heat stress

responses in pigeonpea (*Cajanus cajan*) and its crop wild relatives. Plant Cell Reports. 2021;**40**:881-898

[62] Becklin KM, Anderson JT, Gerhart LM. Examining plant physiological responses to climate change through an evolutionary lens. Plant Physiology. 2016;**172**:639-645

[63] Mishra S, Singh R, Kumar R. Impact of climate change on pigeon pea. Economic Affairs. 2017;**62**:455

[64] Raju BMK, Rao CA, Rao VUM, Rao MS, Maheswari M. Effect of climate on productivity of pigeonpea and cotton in Andhra Pradesh—A panel data regression. Indian Journal of Dryland Agricultural Research and Development. 2014;**29**(1):6-10

[65] Kwena KM, Karuku GN, Ayuke FO, Esilaba AO. Impact of climate change on maize and pigeonpea yields in semi-arid Kenya. In: Environmental Issues and Sustainable Development. London, UK, London, UK: IntechOpen; 2021

[66] O'Dell D, Eash NS, Hicks BB, Oetting JN, Sauer TJ, Lambert DM, et al. Conservation agriculture as a climate change mitigation strategy in Zimbabwe. International Journal of Agricultural Sustainability. 2020;**18**(3):250-265

[67] Grabowski P, Olabisi LS, Adebiyi J, Waldman K, Richardson R, Rusinamhodzi L, et al. Assessing adoption potential in a risky environment: The case of perennial pigeonpea. Agricultural Systems. 2019;**171**:89-99

[68] Sreeharsha RV, Mudalkar S, Sengupta D, Unnikrishnan DK, Reddy AR. Mitigation of drought-induced oxidative damage by enhanced carbon assimilation and an efficient antioxidative metabolism under high CO_2 environment in pigeonpea (*Cajanus cajan* L.). Photosynthesis Research. 2019;**139**(1-3):425-439

[69] Bhagat AA, Tirmali AM, Chaudhary PM. Precipitation distribution and impact of weather parameters on area, production and productivity of pigeonpea in Jalgaon district. International Journal of Chemical Studies. 2020;**8**(4):2167-2172

[70] Yadav MK, Patel C, Singh RS, Singh KK, Balasubramanian R, Mall RK, et al. Assessment of climate change impact on different pigeonpea maturity groups in north Indian condition. Journal of Agrometeorology. 2021;**23**(1):82-92

[71] Bita CE, Gerats T. Plant tolerance to high temperature in a changing environment: Scientific fundamentals and production of heat stress-tolerant crops. Frontiers in Plant Science. 2013;**4**

[72] Kumar S, Saxena VRK, Jackson SA. The Pigeon Pea Genome. Switzerland: Springer International Publishing AG; 2017. pp. 41-54

[73] Bauchet GJ, Bett KE, Cameron CT, Campbell JD, Cannon EK, Cannon SB, et al. The future of legume genetic data resources: Challenges, opportunities, and priorities. Legume Science. 2019;**1**:e16

[74] Gonçalves-Vidigal MC, Gilio TAS, Valentini G, Vaz-Bisneta M, Vidigal Filho PS, Song Q, et al. New Andean source of resistance to anthracnose and angular leaf spot: Fine-mapping of disease-resistance genes in California Dark Red Kidney common bean cultivar. PLoS One. 2020;**15**:e0235215

[75] Chandrashekharaiah PS, Paul V, Kushwaha S, Sanyal D, Dasgupta S. Biotechnological approaches for enhancing stress tolerance in legumes. In: Sustainable Agriculture Reviews. Vol. 51. Cham: Springer; 2021. pp. 247-293

[76] Sekhar K, Priyanka B, Reddy VD, Rao KV. Isolation and characterization of a pigeonpea cyclophilin (CcCYP) gene, and its over-expression in Arabidopsis confers multiple abiotic stress tolerance. Plant, Cell and Environment. 2010;**33**(8):1324-1338

[77] Mellacheruvu S, Tamirisa S, Vudem DR, Khareedu VR. Pigeonpea hybrid-proline-rich protein (CcHyPRP) confers biotic and abiotic stress tolerance in transgenic rice. Frontiers in Plant Science. 2016;**6**:1167

[78] Sunitha M, Srinath T, Reddy VD, Rao KV. Expression of cold and drought regulatory protein (CcCDR) of pigeonpea imparts enhanced tolerance to major abiotic stresses in transgenic rice plants. Planta. 2017;**245**(6):1137-1148

[79] Younis A, Ramzan F, Ramzan Y, Zulfiqar F, Ahsan M, Lim KB. Molecular markers improve abiotic stress tolerance in crops: A review. Plants. 2020;**9**(10):1374

[80] Dong B, Niu L, Meng D, Song Z, Wang L, Jian Y, et al. Genome-wide analysis of MATE transporters and response to metal stress in *Cajanus cajan*. Journal of Plant Interactions. 2019;**14**:265-275

[81] Xie C, H, He W, Hang Y, Li H, Jing Y, Yao R, et al. A high-density SNP genotyping array for rice biology and molecular breeding. Molecular Plant. 2013;7:541-553. DOI: 10.1093/mp/sst135

[82] Singh S et al. Genome-wide distribution, organisation and functional characterization of disease resistance and defence response genes across rice species. PLoS One. 2015;**10**(4):e0125964

[83] Singh S, Mahato AK, Jayaswal PK, et al. A 62K genic-SNP chip array for genetic studies and breeding applications in pigeonpea (*Cajanus cajan* L Millsp). Scientific Reports. 2020;**10**:4960

[84] Varshney RK, Pandey MK, Bohra A, Singh VK, Thudi M. Toward the sequence-based breeding in legumes in the postgenome sequencing era. Theoretical and Applied Genetics. 2018;**132**:797-816

[85] Dinesh HB, Chandappa L, Vishwanath KP, Singh P, Manjunatha L, Ambika DS, et al. Genetic analysis and marker assisted backcrossing for transfer of mosaic virus resistance in cowpea [*Vigna unguiculata* (L.) Walp.]. Legume Research. 2018;**41**:663-668

[86] De Almeida CP, de Carvalho Paulino JF, Bonfante GFJ, Perseguini JMKC, Santos IL, Gonçalves JGR, et al. Angular leaf spot resistance loci associated with different plant growth stages in common bean. Frontiers in Plant Science. 2021;**13**:647043

[87] Binagwa PH, Traore SM, Egnin M, Bernard GC, Ritte I, Mortley D, et al. Genome-wide identification of powdery mildew resistance in common bean (*Phaseolus vulgaris* L.). Frontiers in Genetics. 2021;**22**:69

[88] Nabi A, Lateef I, Nisa Q, Banoo A, Rasool RS, Shah MD, et al. *Phaseolus vulgaris-Colletotrichum lindemuthianum* pathosystem in the post-genomic era: An update. Current Microbiology. 2022;**79**:36

[89] Saxena RK, Hake A, Hingane AJ, Kumar CVS, Bohra A, Sonnappa M, et al. Translational pigeonpea genomics consortium for accelerating genetic gains in pigeonpea (*Cajanus cajan* L.). Agronomy. 2020;**10**:1289

[90] Bohra A, Mir RR, Jha R, Maurya AK, Varshney RK. Advances in genomics and molecular breeding for legume

improvement. In: Narendra T, Renu T, Nishat P, Shabnam KS, editors. Advancement in Crop Improvement Techniques. Kidlington, UK: Woodhead Publishing; 2020. pp. 129-139

[91] Singh G, Singh I, Taggar GK, Rani U, Sharma P, Gupta M, et al. Introgression of productivity enhancing traits, resistance to pod borer and Phytopthora stem blight from Cajanus scarabaeoides to cultivated pigeonpea. Physiology and Molecular Biology of Plants. 2020;**26**:1399-1410

[92] Bohra A, Jha R, Pandey G, Patil PG, Saxena RK, Singh IP, et al. New hypervariable SSR markers for diversity analysis, hybrid purity testing and trait mapping in pigeonpea [*Cajanus cajan* (L.) Millspaugh]. Frontiers in Plant Science. 2017;**8**:1-15

[93] Varshney RK, Close TJ, Singh NK, Hoisington DA, Cook DR. Orphan legume crops enter the genomics era! Current Opinion in Plant Biology. 2009;**12**(2):202-210

[94] Prabha A, Avinashe H, Dubey N. Collection of genomic tools for pigeon pea crop improvement. 2021;**10**(5):324-333

[95] Jha UC, Bohra A, Jha R, Parida SK. Salinity stress response and 'omics' approaches for improving salinity stress tolerance in major grain legumes. Plant Cell Reports. 2019;**38**:255-277

[96] Raju NL, Gnanesh N, Lekha P, Jayashree B, Pande S, Hiremath PJ. The first set of EST resource for gene discovery and marker development in pigeon pea (*Cajanus cajan* L.). BMC Plant Biology. 2010;**10**:45

[97] Dubey A, Farmer A, Schlueter J, Cannon SB, Abernathy B, Tuteja R. Defining the transcriptome assembly and its use for genome dynamics and transcriptome profiling studies in pigeon pea (*Cajanus cajan* L.). DNA Research. 2011;**18**:153-164

[98] Dutta S, Kumawat G, Singh BP, Gupta DK, Singh S, Dogra V. Development of genic-SSR markers by deep transcriptome sequencing in pigeon pea [*Cajanus cajan* (L.) Millspaugh]. BMC Plant Biology. 2011;**11**:17